ILLUSTRATED CHEMISTRY LABORATORY TERMINOLOGY

Herbert W. Ockerman

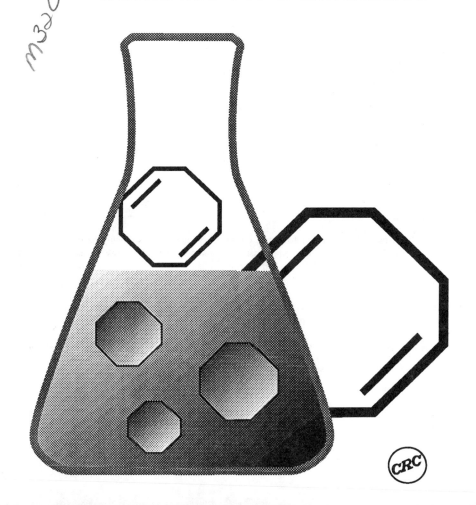

Library of Congress Cataloging-in-Publication Data

Ockerman, Herbert W.
  Illustrated chemistry laboratory terminology : in English, for
speakers of other languages / Herbert W. Ockerman.
      p. cm.
  Includes index.
  ISBN 0-8493-0152-1
  1. Chemical laboratories — Terminology.   I. Title.
QD51.025 1991
542′.1′014 — dc20   91-8591
                                                                           CIP

This book represents information obtained from authentic and highly regarded sources. Re-printed material is quoted with permission, and sources are indicated. Every reasonable effort has been made to give reliable data and information, but the author and the publisher cannot assume responsibility for the validity of all materials or for the consequences of their use.

Direct all inquiries to CRC Press, Inc., 2000 Corporate Blvd., N.W., Boca Raton, Florida, 33431.

International Standard Book Number 0-8493-0152-1

Library of Congress Card Number 91-8591
Printed in the United States

DEDICATED TO FRANCES

Drawings, pictures and photographs were graciously contributed by Thomas Scientific Company and the Fisher Scientific Company. These figures are keyed with either a

$\boxed{\text{T}}$ (Thomas) or a

$\boxed{\text{F}}$ (Fisher).

If you desire to locate these items information on them may be obtained as follows:

Arthur H. Thomas Company
99 High Hill Road at I-295, P. O. Box 99
Swedesboro, N. J. 08085-0099, U. S. A.
Telephone: 609-467-2000
Fax: 609-467-3087
TWX: 710-991-8749
Cable address: BALANCE, Swedesboro

Fisher Scientific Company
50 Fadem Road
Springfield, N. J., 07081
Telephone: 201-467-6400
Fax: 201-379-7415
Telex: 475-4246 or 138287
Cable address: Fishersci, Springfield, N. J.

Permission of these scientific supply companies to reproduce these figures made the publication of this manuscript much easier and the author is grateful.

The author also wishes to thank the many international scientists, translators and illustrators who aided in checking, and translating and illustrating this text.

Scientist or Translator	Affiliation	Area
Dr. R.L.S. Patterson	Agricultural Research Council, Meat Research Institute, Langford, Bristol, England	British English
Mr. Jeng Chii-Yeng	Department of Animal Science, The Ohio State University	Cartoon Illustrations
Dr. Ming-Tsao Chen	National Chung-Hsing University, Taiwan, R.O.C.	Chinese
Dr. Francoise Watts	Columbus, Ohio	French
Mr. Chris Mirski	German Department The Ohio State University	German
Dr. Jacek Szczawinski	Faculty of Veterinary Science, Agriculture University of Warsaw, Warsaw, Poland	Polish
Dr. Francisco Crespo Leon	Facultad de Veterinaria Departamento de Tecnologia y Bioquimica de Los Alimentos, Cordoba, Spain	Spanish
Mr. Hasan Yetim	Tarim Urunleri Teknolojisi Bolumu Ataturk University 25240 Erzurum, Turkey	Turkish
Mr. Robert R. Hodges	Department of Animal Science, The Ohio State University	Computer

PURPOSE

This manual is intended for chemists whose first language is not English and who are attempting to expand their English vocabulary in the chemical laboratory area. It will be of most use to students who have a moderate knowledge of the English language and who have had some practical experience in a chemistry laboratory. It should be extremely useful to non-English-speaking students who plan to teach chemistry (in English) or who are required to write in English a thesis or dissertation on a chemically-related topic.

Items in parentheses indicate British English term or British English spelling or American English alternate term or American English alternate spelling.

Suggestions for improvement of this book are welcomed by the author.

Herbert W. Ockerman, Ph.D.
Professor

SAFETY EQUIPMENT **1**
First Aid Kit 1
Fire Extinguisher 1
Fire Blanket 1
Shower 1
Eye Wash 2
Telephone (Phone) 2
Warning Labels, Warning Signs 2
caution 2
explosive 2
biohazard 2
radiation hazard 3
wear your goggles 3
danger 3
flammable 3
corrosive 3
poison 3
use in hood 3
under high pressure 3
under low pressure 3
caustic 3
avoid contact with skin 3
hot 3
sterile 3
keep in refrigerator 3
out of order 3
Safety Goggles 4
Respirator 4
Discussion 5
LABORATORY FURNITURE **6**
Fume Hood (Fume cupboard) 6
Base Unit 6
Storage Cabinet 6
Floor Case 7
Desk 7
Table 7
Filing Cabinet 7
Discussion 8
PORTABLE LABORATORY FURNITURE **9**
Laboratory Cart (Laboratory trolley) 9
Laboratory Chair or Stool 9
Ladder (Steps) 9
Discussion 10
FIXTURES **11**
Faucet (Tap) 11
Mixing Faucet (Mixer tap) 11
Sink 11
Stopcock 12
Cold Water 12
Hot Water 12
Distilled Water 12
Steam 12
Air 12

FIXTURES (Cont'd)
Gas 13
Oxygen (O_2) 13
Nitrogen (N_2) 13
Vacuum 13
Needle Valve 13
Siphon 13
Discussion 14
ELECTRICAL **15**
Electrical Outlet (Electrical sockets) . . . 15
Outlet Box or Strip (Distribution board or
    box) 15
Variable Transformer or Powerstat ("Variac"). 15
Battery; Chargeable and Non-chargeable . . . 16
Battery Charger 16
Discussion 17
CYLINDER GAS **18**
Gas Cylinder or Tank 18
Pressure Regulator 18
Gas Cylinder Clamp 18
Discussion 19
VACUUM AND AIR PUMPS **20**
Vacuum and Pressure Pump 20
Vane Pump Movement 20
Vacuum Pump (Rotary oil pump) 20
Hand Pump 21
Hand Pump, Floor Model 21
Aspirator (Water pump) 21
Discussion 22
COLD STORAGE **23**
Refrigerator-Freezer 23
Dry Ice Storage Container (Solid carbon
    dioxide storage container) 23
Dry Ice Box (Solid carbon dioxide storage or
    shipping container) 24
Dewar Flask (Thermos flask) 24
Discussion 25
HAND LABORATORY EQUIPMENT **26**
Spatula 26
Scoop 26
Tongs 26
Policeman 26
Notebook 26
Foil 27
Bag 27
Evaporating Dish 27
Apron 27
Asbestos Gloves 28
Rubber Gloves 28
Disposable Gloves 28
Wash Bottle 28
Dropping Bottle 29
Rubber Bulb 29
Brush 29
Cleaning Compound 29

HAND LABORATORY EQUIPMENT (Cont'd)
Tissue 29
Wall Clock 30
Interval Timer 30
Time Switch 30
Stopwatch 31
Discussion 32
LABORATORY HAND TOOLS **33**
Pliers 33
Needle Nose Pliers (Fine nose pliers) 33
Wire Cutter Pliers 33
Tin Snips 33
Shears or Scissors 33
Crescent Wrench (Adjustable wrench) 34
Open-end Wrench 34
Hex Wrench Set 34
Hammer 34
Screwdriver 34
Drill Bit 35
Metal File 35
Vise (Vice) 35
Soldering Gun (Soldering iron) 35
Discussion 36
OFFICE HAND TOOLS **37**
Tape Dispenser 37
Label Maker 37
Label 37
Ball Point Pen 37
Wax Pencil 38
Felt Tip Marking Pen 38
Rule or Ruler 38
Calculator 38
Discussion 39
PERIODIC TABLE **40**
Periodic Table or Chart 40
Discussion 41
REAGENT **42**
Acetic Acid 42
Alcohol 42
Ammonium Hydroxide 42
Calcium Hydroxide 42
Hydrochloric Acid 42
Nitric Acid 42
Potassium Hydroxide 42
Sodium Hydroxide 42
Sulfuric Acid (Sulphuric acid) 42
Reagent Label 43
Discussion 44
MOLECULAR MODELS **45**
SUPPORTS **46**
Support or Frame 46
Buret Stand (Burette stand) 46
Ring Stand (Retort stand) 46
Tripod 47
Buret Clamp (Burette clamp) 47

SUPPORTS (Cont'd)
Clamp Holder   47
Clamp   47
Clamp with Holder   48
Discussion   49
GLASS **50**
Glass Tubing   50
Capillary Tubing   50
Glass Rod   50
Ball and Socket Joint   51
O-Ring Joint   51
Standard Taper Ground Glass Joint ("Quickfit")   51
Pinch Clamp   51
Glass Stopcock   52
Condenser   52
Test Tube   52
Test Tube Holder   53
Test Tube Rack   53
Watch Glass   53
Safety Hand Grip   53
Glass Tubing Cutter   54
Discussion   55
CONTAINERS AND CLOSURES **56**
Beaker   56
Bottle   56
Erlenmeyer Flask (Conical flask)   56
Pail or Bucket   56
Florence Flask   57
Vial   57
Cap   57
Stopper; Rubber or Plastic Stopper   57
Glass Stopper   57
Cork   58
Cork Borer   58
Discussion   59
TUBING **60**
Rubber Tubing   60
Plastic Tubing   60
Tubing Clamp   60
Discussion   61
LIQUID VOLUME MEASURING EQUIPMENT **62**
Syringe Pipet (Syringe pipette)   62
Transfer or Volumetric Pipet (Transfer or
    Volumetric pipette)   62
Measuring or Graduated Pipet (Measuring or
    Graduated pipette)   62
Pipet Filling Attachment or Bulb (Pipette
    Filling attachment or bulb)   63
Automatic Transfer Pipet (Automatic transfer
    pipette)   63
Graduated Cylinder   63
Graduate   64
Beaker   64
Graduated Buret (Graduated burette)   64
Automatic Zero Pipet (Automatic zero pipette)   65

LIQUID VOLUME MEASURING EQUIPMENT (Cont'd)
Stopcock 65
Stopcock Grease 65
Volumetric Flask 66
Pump 66
Discussion 67
WEIGHT AND SAMPLE COMPOSITION **68**
Scale 68
Triple Beam Balance 68
Balance 68
Top-Loading Balance 69
Weight Set 69
Specific Gravity Balance 69
Hydrometer 70
Specific Gravity Bottle; Pycnometer 70
Viscometer 70
Polarimeter 71
Refractometer 71
Hand Refractometer 71
Discussion 72
HEATING **73**
Oven 73
Hot Plate 73
Incubator 73
Heating Mantle 74
Immersion Heater 74
Heating Tape 74
Muffle Furnace 75
Crucible 75
Water Bath 75
Thermometer 76
Dial Thermometer 76
Thermometer (Electronic) 76
Autoclave or Sterilizer 77
Instrument Sterilizer 77
Burner 77
Alcohol Lamp 78
Gas Lighter 78
Safety Matches 78
Melting Point Bath 79
Melting Point Tube 79
Discussion 80
MIXING AND PRESSING EQUIPMENT **81**
Mortar and Pestle 81
Stirrer 81
Magnetic Stirrer 81
Large Size Shaker 82
Shaker ("Whirli mixer") 82
Blender 82
Hand Grinder (Hand mincer) 83
Mill 83
Tissue Grinder 83
Homogenizer 84
Emulsion Homogenizer 84
Hydraulic Press 85

MIXING AND PRESSING EQUIPMENT (Cont'd)
Discussion   86
MEASUREMENT EQUIPMENT **87**
pH Meter   87
Electrode   87
pH Buffer Solution   87
Test Paper or Litmus Paper or Universal
    Indicator Paper   88
Volt-Ohm-Ammeter (AVO-meter)   88
Barometer   88
Bomb Calorimeter   89
Centrifuge   89
Centrifuge Head (Rotor)   90
Centrifuge Tube   90
Carriers   90
Spectrophotometer   91
Spectro Cell   91
Fluorometer (Fluorimeter)   92
Flowmeter   92
Discussion   93
SEPARATION EQUIPMENT **94**
Gel Column Electrophoresis   94
Scanning Densitometer   94
Dialyzer   94
Automatic Sampler   95
Sieve   95
Separatory Funnel (Separating funnel)   95
Chromatography Column   96
Thin Layer Chromatography (TLC)   96
Chromatography Cabinet   96
Soxhlet Extractor   97
Extraction Thimble   97
Goldfisch Extractor   97
Babcock Bottle   98
Divider   98
Babcock Centrifuge   98
Aluminum Foil Dish (Aluminium foil dish) . .   99
Desiccator   99
Drying Oven   99
Vacuum Oven   100
Moisture Balance   100
Freeze Drying   100
Kjeldahl Flask   101
Kjeldahl Digestion Unit   101
Kjeldahl Distilling Unit   101
Recorder   102
Discussion   103
FILTERING **104**
Funnel   104
Filter Paper   104
Glass Wool   104
Filtering Flask (Filter flask)   105
Buchner Filter Funnel   105
Crucible   105
Crucible Holder   106

FILTERING (Cont'd)
Impinger 106
Discussion 107
DISTILLATION **108**
Still 108
Distilling Flask (Distillation flask) 108
Water Still 108
Demineralizer 109
Flash Evaporator 109
Discussion 110
GAS MEASUREMENT **111**
Respirometer 111
Discussion 112
SMALL ANIMAL EQUIPMENT **113**
Animal Cage 113
Drinking Tube 113
Feeding Dish 113
Dissecting Kit 114
Animal Operating Table 114
Discussion 115
MICROSCOPIC EXAMINATION **116**
Microscope 116
Slide 116
Magnifier 116
Lens Tissue 117
Micrometer 117
Counting Chamber 117
Staining Dish or Jar 118
Tissue Processing Unit 118
Microtome 118
Discussion 119
MICROBIOLOGY **120**
Culture Media 120
Dilution Bottle 120
Petri Dish 120
Petri Dish Sterilizing Box 121
Quebec Colony Counter 121
Counter 121
Inoculating Needle or Loop 122
Culture Tube 122
Discussion 123
WASHING **124**
Glassware Washer 124
Pipet Washer (Pipette washer) 124
Draining Rack 125
Discussion 126

中 文 目 錄

安全設施···································· 1

急救箱····································· 1

滅火器····································· 1

防火氈····································· 1

蓮蓬頭（噴水器）···························· 1

眼睛沖洗具································· 2

電　話···································· 2

警告標識、警告記號························· 2

注意、小心······························· 2

爆炸性·································· 2

生物危險性······························· 2

放射性危險性····························· 3

帶上護目鏡······························· 3

危　險································· 3

易燃的································· 3

腐蝕性································· 3

有　毒································· 3

在毒氣室內使用························· 3

在高壓下······························· 3

在低壓下······························· 3

腐蝕性································· 3

避免與皮膚接觸························· 3

熱·································· 3

無　菌································ 3

保存在冰箱···························· 3

壞　掉······························· 3

安全護目眼鏡 · · · · · · · · · · · · · · · 　4

面罩；呼吸器、濾毒罐 · · · · · · · · · · · 　4

討　論 · · · · · · · · · · · · · · · · · · 　5

實驗室器具 · · · · · · · · · · · · · · · · · 　6

烟櫥；通風櫥、抽油煙機 · · · · · · · · · · 　6

基本單位 · · · · · · · · · · · · · · · · · 　6

儲存櫥（箱、小室） · · · · · · · · · · · · 　6

落地櫥 · · · · · · · · · · · · · · · · · · 　7

實驗檯 · · · · · · · · · · · · · · · · · · 　7

桌子；實驗桌 · · · · · · · · · · · · · · · 　7

銼屑櫥 · · · · · · · · · · · · · · · · · · 　7

討　論 · · · · · · · · · · · · · · · · · · 　8

輕便的（可移動的）實驗室器具 · · · · · · · · 　9

實驗室吊運車 · · · · · · · · · · · · · · · 　9

實驗室椅子或凳子 · · · · · · · · · · · · · 　9

梯　子 · · · · · · · · · · · · · · · · · · 　9

討　論 · · · · · · · · · · · · · · · · · · 　10

裝置品 · · · · · · · · · · · · · · · · · · · 　11

水龍頭（塞子；栓；龍頭） · · · · · · · · · 　11

混合水龍頭（混合器龍頭） · · · · · · · · · 　11

水槽；水斗、漏水池 · · · · · · · · · · · · 　11

轉門；（活）栓、活塞 · · · · · · · · · · · 　12

冷　水 · · · · · · · · · · · · · · · · · · 　12

熱　水 · · · · · · · · · · · · · · · · · · 　12

蒸餾水 · · · · · · · · · · · · · · · · · · 　12

蒸汽、水蒸汽 · · · · · · · · · · · · · · · 　12

空　氣 · · · · · · · · · · · · · · · · · · 　12

氣體；氣態；煤氣；毒氣 · · · · · · · · · · 　13

氧；氧氣 · 13

氮；氮氣 · 13

真　空 · 13

針孔閥 · 13

虹吸、虹吸管 · 13

討　論 · 14

電　器 · 15

電的出口管；插座 · · · · · · · · · · · · · · · · · · · 15

插座箱；配線箱 · 15

可變的變壓器 · 15

電池組（可充電的和不可充電的） · · · · · · · · · · · 15

電池充電器 · 16

討　論 · 17

汽缸氣、氣體汽缸（鋼瓶圓筒） · · · · · · · · · · · · · 18

氣體鋼瓶；汽缸（箱） · · · · · · · · · · · · · · · · · 18

調壓器 · 18

汽缸夾（壓板） · 18

討　論 · 19

真空空氣唧筒（泵浦） · · · · · · · · · · · · · · · · · 20

真空壓氣泵浦 · 20

葉輪式可移動泵 · 20

真空泵；輪迴抽油機 · · · · · · · · · · · · · · · · · · 20

手動抽水機（泵） · · · · · · · · · · · · · · · · · · · 21

地板式手動抽水機（泵） · · · · · · · · · · · · · · · · 21

吸氣器（水泵） · 21

討　論 · 21

冷　藏 · 22

冷凍機—冷藏箱 · 23

乾冰貯存器（固體二氧化碳貯存器） · · · · · · · · · · · 23

乾冰盒（固體二氧化碳貯存或運輸）容器 · · · · · · · · · 24

真空瓶；杜瓦瓶（保溫瓶） · · · · · · · · · · · · · · · 24

討　論 · 25

實驗室使用設備 · · · · · · · · · · · · · · · · · · · 26

　刮鏟、鴨舌匙 · · · · · · · · · · · · · · · · · 26

　勺、杓、挖空 · · · · · · · · · · · · · · · · · 26

　鉗　子 · 26

　澱　帚 · 26

　筆記簿 · 26

　錫箔片 · 27

　袋　子 · 27

　蒸發皿 · 27

　圍　裙 · 27

　石棉手套 · · · · · · · · · · · · · · · · · · · 28

　橡膠手套 · · · · · · · · · · · · · · · · · · · 28

　用後可丟棄手套 · · · · · · · · · · · · · · · · 28

　洗滌瓶 · 28

　滴　瓶 · 29

　橡膠球管（球） · · · · · · · · · · · · · · · · 29

　刷　子 · 29

　洗滌劑 · 29

　擦拭紙 · 29

　壁　鐘 · 30

　間隔計時器 · · · · · · · · · · · · · · · · · · 30

　定時轉變器 · · · · · · · · · · · · · · · · · · 30

　碼錶、計時錶 · · · · · · · · · · · · · · · · · 31

　討　論 · 32

實驗室使用工具 · · · · · · · · · · · · · · · · · · · 33

　鉗　子 · · · · · · · · · · · · · · · · · · · 33

　針孔鉗（細孔鉗）· · · · · · · · · · · · · 33

　鋼絲鉗 · · · · · · · · · · · · · · · · · · · 33

　剪錫鉗 · · · · · · · · · · · · · · · · · · · 33

　大剪刀、剪刀 · · · · · · · · · · · · · · · 33

　新月形扳鉗（可調節扳剪）· · · · · · · · 34

　開端扳鉗 · · · · · · · · · · · · · · · · · · 34

　六扳鉗組 · · · · · · · · · · · · · · · · · · 34

　錘 · 34

　螺絲起子；螺絲刀 · · · · · · · · · · · · · 34

　螺旋鑽 · · · · · · · · · · · · · · · · · · · 35

　金屬銼刀 · · · · · · · · · · · · · · · · · · 35

　虎頭鉗 · · · · · · · · · · · · · · · · · · · 35

　焊接槍 · · · · · · · · · · · · · · · · · · · 35

　討　論 · · · · · · · · · · · · · · · · · · · 36

辦公室使用工具 · · · · · · · · · · · · · · · · · · 37

　紮帶分配器 · · · · · · · · · · · · · · · · · 37

　標籤製造器 · · · · · · · · · · · · · · · · · 37

　標籤 · 37

　原子筆 · · · · · · · · · · · · · · · · · · · 37

　蠟筆 · 38

　簽字筆 · · · · · · · · · · · · · · · · · · · 38

　尺；米達尺 · · · · · · · · · · · · · · · · · 38

　計算機 · · · · · · · · · · · · · · · · · · · 38

　討　論 · · · · · · · · · · · · · · · · · · · 39

週期表・・・・・・・・・・・・・・・・・・・・・・ 40

　週期表（或圖）・・・・・・・・・・・・・・・・ 40

　討　論・・・・・・・・・・・・・・・・・・・・ 41

試　藥　・・・・・・・・・・・・・・・・・・・・・ 42

　醋　酸・・・・・・・・・・・・・・・・・・・・ 42

　酒　精・・・・・・・・・・・・・・・・・・・・ 42

　氫氧化銨・・・・・・・・・・・・・・・・・・・ 42

　氫氧化鈣・・・・・・・・・・・・・・・・・・・ 42

　氫氯酸；塩酸・・・・・・・・・・・・・・・・・ 42

　硝　酸・・・・・・・・・・・・・・・・・・・・ 42

　氫氧化鉀・・・・・・・・・・・・・・・・・・・ 42

　氫氧化鈉・・・・・・・・・・・・・・・・・・・ 42

　硫　酸・・・・・・・・・・・・・・・・・・・・ 42

　試藥標籤・・・・・・・・・・・・・・・・・・・ 43

　討　論・・・・・・・・・・・・・・・・・・・・ 44

分子模型・・・・・・・・・・・・・・・・・・・・・ 45

架；支座・・・・・・・・・・・・・・・・・・・・・ 46

　架；支座或框・・・・・・・・・・・・・・・・・ 46

　滴定管架・・・・・・・・・・・・・・・・・・・ 46

　環架（蒸餾瓶架）・・・・・・・・・・・・・・・ 46

　三脚架・・・・・・・・・・・・・・・・・・・・ 47

　滴定管夾・・・・・・・・・・・・・・・・・・・ 47

　夾子固定器・・・・・・・・・・・・・・・・・・ 47

　夾　子・・・・・・・・・・・・・・・・・・・・ 47

　固定夾・・・・・・・・・・・・・・・・・・・・ 48

　討　論・・・・・・・・・・・・・・・・・・・・ 49

玻璃器具 · 50

  玻璃管 · · · · · · · · · · · · · · · · · · · 50

  毛細管 · · · · · · · · · · · · · · · · · · · 50

  玻璃棒 · · · · · · · · · · · · · · · · · · · 50

  球和套筒聯接 · · · · · · · · · · · · · · · 51

  O 型環聯接 · · · · · · · · · · · · · · · · 51

  標準錐形磨口玻璃接頭 · · · · · · · · · 51

  揑　夾 · · · · · · · · · · · · · · · · · · · 51

  玻璃栓（轉閂） · · · · · · · · · · · · · · 52

  冷凝器、電容器 · · · · · · · · · · · · · · 52

  試　管 · · · · · · · · · · · · · · · · · · · 52

  試管夾 · · · · · · · · · · · · · · · · · · · 53

  試管架 · · · · · · · · · · · · · · · · · · · 53

  表玻璃 · · · · · · · · · · · · · · · · · · · 53

  安全手柄 · · · · · · · · · · · · · · · · · · 53

  玻璃管切斷機 · · · · · · · · · · · · · · · 54

  討　論 · · · · · · · · · · · · · · · · · · · 55

容器和密封器 · · · · · · · · · · · · · · · · · · · 56

  燒　杯 · · · · · · · · · · · · · · · · · · · 56

  瓶　子 · · · · · · · · · · · · · · · · · · · 56

  錐瓶、錐形燒瓶或三角瓶 · · · · · · · · 56

  桶；吊桶；斗 · · · · · · · · · · · · · · · 56

  平底燒瓶 · · · · · · · · · · · · · · · · · · 57

  管形瓶 · · · · · · · · · · · · · · · · · · · 57

  蓋、帽、罩 · · · · · · · · · · · · · · · · · 57

容器和密閉器（續）

　塞子；橡膠或塑膠塞子 · · · · · · · · · · · 57

　玻璃塞子 · · · · · · · · · · · · · · · · 57

　軟木塞 · · · · · · · · · · · · · · · · · 58

　塞子穿孔器 · · · · · · · · · · · · · · · 58

　討　論 · · · · · · · · · · · · · · · · · 59

試管：管子 · · · · · · · · · · · · · · · · · 60

　橡膠試管 · · · · · · · · · · · · · · · · 60

　塑膠試管 · · · · · · · · · · · · · · · · 60

　試管夾 · · · · · · · · · · · · · · · · · 60

　討　論 · · · · · · · · · · · · · · · · · 61

液體體積測量設備 · · · · · · · · · · · · · · 62

　注射吸管（注射筒） · · · · · · · · · · · 62

　可移動的定量吸管 · · · · · · · · · · · · 62

　測量或有刻度吸管 · · · · · · · · · · · · 62

　有吸球吸管 · · · · · · · · · · · · · · · 63

　自動轉移吸管 · · · · · · · · · · · · · · 63

　量　筒 · · · · · · · · · · · · · · · · · 63

　量　杯 · · · · · · · · · · · · · · · · · 64

　燒　杯 · · · · · · · · · · · · · · · · · 64

　有刻度定量管（滴定管） · · · · · · · · · 64

　自動歸零吸管 · · · · · · · · · · · · · · 65

　栓；轉閂 · · · · · · · · · · · · · · · · 65

　活栓脂膏 · · · · · · · · · · · · · · · · 65

　定量瓶 · · · · · · · · · · · · · · · · · 66

　泵、唧筒 · · · · · · · · · · · · · · · · 66

　討　論 · · · · · · · · · · · · · · · · · 67

加　熱（續）

　　高壓加熱器；壓熱器、高壓殺菌器・・・・・・・・・・・・・・　77

　　殺菌器；消毒器・・・・・・・・・・・・・・・・・・・・　77

　　燈頭；燃燒器・・・・・・・・・・・・・・・・・・・・・　77

　　酒精燈・・・・・・・・・・・・・・・・・・・・・・・・　78

　　瓦斯（氣體）點火器（發光器）・・・・・・・・・・・・・　78

　　安全火柴・・・・・・・・・・・・・・・・・・・・・・・　78

　　熔點浴・・・・・・・・・・・・・・・・・・・・・・・・　79

　　熔點管・・・・・・・・・・・・・・・・・・・・・・・・　79

　　討　論・・・・・・・・・・・・・・・・・・・・・・・・　80

混合和加壓設備・・・・・・・・・・・・・・・・・・・・・・　81

　　白和杵・・・・・・・・・・・・・・・・・・・・・・・・　81

　　攪拌器・・・・・・・・・・・・・・・・・・・・・・・・　81

　　磁性攪拌器・・・・・・・・・・・・・・・・・・・・・・　81

　　大號搖動器・・・・・・・・・・・・・・・・・・・・・・　82

　　搖動器（迴轉混合器）・・・・・・・・・・・・・・・・・　82

　　摻合器・・・・・・・・・・・・・・・・・・・・・・・・　82

　　手動碾磨機（手動絞碎機）・・・・・・・・・・・・・・・　83

　　磨碾機・・・・・・・・・・・・・・・・・・・・・・・・　83

　　組織碾磨機・・・・・・・・・・・・・・・・・・・・・・　83

　　均質機・・・・・・・・・・・・・・・・・・・・・・・・　84

　　乳濁液均質機・・・・・・・・・・・・・・・・・・・・・　84

　　水壓機・・・・・・・・・・・・・・・・・・・・・・・・　85

　　討　論・・・・・・・・・・・・・・・・・・・・・・・・　86

測量設備・・・・・・・・・・・・・・・・・・・・・・・　87

　PH　測定器、PH　計・・・・・・・・・・・・　87

　電　　極・・・・・・・・・・・・・・・・・・・・・・　87

　PH　（酸、鹼）緩衝液・・・・・・・・・・・・・　87

　試管或石蕊試紙或通用試紙・・・・・・・・・・　88

　伏特－歐姆安培計・・・・・・・・・・・・・・・　88

　氣壓計・・・・・・・・・・・・・・・・・・・・・・　88

　彈式量熱器（熱量測定器）・・・・・・・・・・　89

　離心機・・・・・・・・・・・・・・・・・・・・・・　89

　離心機轉頭・・・・・・・・・・・・・・・・・・・　90

　離心管・・・・・・・・・・・・・・・・・・・・・・　90

　載體、遞體・・・・・・・・・・・・・・・・・・・　90

　分光光度計・・・・・・・・・・・・・・・・・・・　91

　光譜室（箱）・・・・・・・・・・・・・・・・・・　91

　氣量計；螢光計・・・・・・・・・・・・・・・・・　92

　流量計・・・・・・・・・・・・・・・・・・・・・・　92

　討　　論・・・・・・・・・・・・・・・・・・・・・・　93

分離設備・・・・・・・・・・・・・・・・・・・・・・・　94

　凝膠柱電泳・・・・・・・・・・・・・・・・・・・　94

　螢光密度計・・・・・・・・・・・・・・・・・・・　94

　滲析膜；滲析器・・・・・・・・・・・・・・・・・　94

　自動採樣器・・・・・・・・・・・・・・・・・・・　95

　篩　　子・・・・・・・・・・・・・・・・・・・・・・　95

　分液漏斗・・・・・・・・・・・・・・・・・・・・・　95

　色層吸附柱；色譜柱・・・・・・・・・・・・・・　96

　薄層色層分析法（TLC）・・・・・・・・・・・　96

　色層分析櫥（箱）・・・・・・・・・・・・・・・・　96

分離設備（續）

索格利特萃取器 • • • • • • • • • • • • • • 97

提取殼筒 • • • • • • • • • • • • • • • • 97

Goldfisch 萃取器 • • • • • • • • • • • • 97

巴布科克乳油管 • • • • • • • • • • • • • 98

分規；雙腳規 • • • • • • • • • • • • • 98

巴布科克遠心分離器 • • • • • • • • • • • 98

鋁箔皿（碟） • • • • • • • • • • • • • • 99

除濕乾燥器 • • • • • • • • • • • • • • • • 99

乾燥烘箱 • • • • • • • • • • • • • • • • 99

真空烘箱 • • • • • • • • • • • • • • • • 100

測濕器（水分測定計） • • • • • • • • • • 100

冷凍乾燥 • • • • • • • • • • • • • • • • 100

長頸燒瓶；凱氏燒瓶 • • • • • • • • • • • 101

凱氏分解裝置 • • • • • • • • • • • • • • 101

凱氏蒸餾裝置 • • • • • • • • • • • • • • 101

記錄器 • • • • • • • • • • • • • • • • • 102

討　論 • • • • • • • • • • • • • • • • • 103

過　濾 • • • • • • • • • • • • • • • • • 104

漏　斗 • • • • • • • • • • • • • • • • • 104

濾　紙 • • • • • • • • • • • • • • • • • 104

玻璃棉；玻璃絨 • • • • • • • • • • • • • 104

吸濾瓶 • • • • • • • • • • • • • • • • • 105

瓷過濾漏斗 • • • • • • • • • • • • • • • 105

坩　堝 • • • • • • • • • • • • • • • • • 105

過　濾（續）

　　坩堝夾・・・・・・・・・・・・・・・・・・・・・・106

　　衝擊板；打擊板・・・・・・・・・・・・・・・・・・106

　　討　論・・・・・・・・・・・・・・・・・・・・・・・107

蒸　餾・・・・・・・・・・・・・・・・・・・・・・・・・108

　　蒸餾釜；蒸餾鍋・・・・・・・・・・・・・・・・・・108

　　蒸餾瓶・・・・・・・・・・・・・・・・・・・・・・・108

　　蒸水器・・・・・・・・・・・・・・・・・・・・・・・108

　　脫礦物質器・・・・・・・・・・・・・・・・・・・・109

　　急驟蒸發器・・・・・・・・・・・・・・・・・・・・109

　　討　論・・・・・・・・・・・・・・・・・・・・・・・110

氣體測量器・・・・・・・・・・・・・・・・・・・・・・111

　　呼吸計・・・・・・・・・・・・・・・・・・・・・・・111

　　討　論・・・・・・・・・・・・・・・・・・・・・・・112

小動物設備・・・・・・・・・・・・・・・・・・・・・・113

　　動物籠（檻）・・・・・・・・・・・・・・・・・・・113

　　飲水管・・・・・・・・・・・・・・・・・・・・・・・114

　　飼養盤（槽）・・・・・・・・・・・・・・・・・・・113

　　解剖用具・・・・・・・・・・・・・・・・・・・・・114

　　動物手術檯・・・・・・・・・・・・・・・・・・・・114

　　討　論・・・・・・・・・・・・・・・・・・・・・・・115

顯微鏡的檢查・・・・・・・・・・・・・・・・・・・・116

　　顯微鏡・・・・・・・・・・・・・・・・・・・・・・・116

　　玻璃片；幻燈片・・・・・・・・・・・・・・・・・・116

　　放大鏡・・・・・・・・・・・・・・・・・・・・・・・116

顯微鏡的檢查（續）

　　鏡頭擦拭紙・・・・・・・・・・・・・・・・・117

　　測微計・・・・・・・・・・・・・・・・・・・117

　　計算室・・・・・・・・・・・・・・・・・・・117

　　染色盤或染色缸・・・・・・・・・・・・・・・118

　　組織處理單位・・・・・・・・・・・・・・・・118

　　切片機・・・・・・・・・・・・・・・・・・・118

　　討　論・・・・・・・・・・・・・・・・・・・119

微生物學・・・・・・・・・・・・・・・・・・・・120

　　培養基・・・・・・・・・・・・・・・・・・・120

　　稀釋瓶・・・・・・・・・・・・・・・・・・・120

　　培養皿・・・・・・・・・・・・・・・・・・・120

　　培養皿殺菌盒・・・・・・・・・・・・・・・・121

　　Quebec 菌落計數器（計群器）・・・・・・・・・121

　　計數器・・・・・・・・・・・・・・・・・・・121

　　接種針或環、白金耳・・・・・・・・・・・・・122

　　培養管・・・・・・・・・・・・・・・・・・・122

　　討　論・・・・・・・・・・・・・・・・・・・123

洗　　滌・・・・・・・・・・・・・・・・・・・・124

　　玻璃儀器洗滌機（器）・・・・・・・・・・・・124

　　吸管洗滌機（器）・・・・・・・・・・・・・・124

　　排水架、滴乾架・・・・・・・・・・・・・・・125

　　討　論・・・・・・・・・・・・・・・・・・・126

MATERIEL DE SECURITE 1
    Trousse de premier secours 1
    Extincteur 1
    Couverture anti-feu 1
    Douche 1
    Lave-oeil 2
    Téléphone 2
    Etiquettes et affiches 2
        danger 2
        explosif 2
        risque biologique 2
        danger de radiations 3
        mettez vos lunettes 3
        danger 3
        inflammable 3
        corrosif 3
        poison 3
        utiliser sous hotte 3
        sous haute pression 3
        sous basse pression 3
        caustique 3
        éviter le contact avec la peau 3
        chaud 3
        stérile 3
        garder au refrigérateur 3
        en panne 3
    Lunettes de sécurité 4
    Masque 4
    Discussion 5
MOBILIER DE LABORATOIRE 6
    Hotte de ventilation 6
    Socle 6
    Placard de rangement 6
    Cabinet au sol 7
    Bureau 7
    Table 7
    Meuble de classement 7
    Discussion 8
MOBILIER DE LABORATOIRE PORTABLE 9
    Charriot de laboratoire 9
    Tabouret de laboratoire 9
    Escabeau 9
    Discussion 10
INSTALLATIONS 11
    Robinet 11
    Robinet-mélangeur d'eau 11
    Evier 11
    Bouchon d'arrêt 12
    Eau froide 12
    Eau chaude 12
    Eau distillée 12
    Vapeur 12

INSTALLATIONS (suite)
    Air 12
    Gaz 13
    Oxygène (O₂) 13
    Nitrogène (N₂) 13
    Vide 13
    Soupape en aiguille 13
    Siphon 13
    Discussion 14
APPAREILLAGE ELECTRIQUE 15
    Prise électrique 15
    Disjoncteur 15
    Transformateur variable 15
    Batterie, chargeable et non-rechargeable . . . 16
    Rechargeur de piles 16
    Discussion 17
BOUTEILLE A GAZ 18
    Bouteille à gaz 18
    Régulateur de pression 18
    Support de bouteille à gaz 18
    Discussion 19
VIDE ET POMPE A AIR 20
    Pompe à vide et à pression 20
    Mouvement de pompe à vide 20
    Pompe à vide 20
    Pompe à main 21
    Pompe à main, modèle au sol 21
    Aspirateur (pompe à eau) 21
    Discussion 22
REFRIGERATION 23
    Refrigérateur-congélateur 23
    Cuve de stockage de glace sèche 23
    Boîte de glace sèche 24
    Bouteille Thermos 24
    Discussion 25
MATERIEL DE LABORATOIRE A MANIEMENT MANUEL 26
    Spatule 26
    Cuillère 26
    Pinces 26
    Agitateur 26
    Carnet 26
    Papier aluminium 27
    Sac en papier 27
    Assiette d'évaporation 27
    Tablier 27
    Gants en amiante 28
    Gants de caoutchouc 28
    Gants à jeter 28
    Flacon laveur 28
    Flacon de compte-gouttes 29
    Ampoule en caoutchouc 29
    Brosse 29
    Produit à nettoyer 29

MATERIEL DE LABORATOIRE A MANIEMENT MANUEL (suite)
      Papier-mouchoir   29
      Horloge murale   30
      Minuteur   30
      Interrupteur automatique   30
      Chronomètre   31
      Discussion   32
INSTRUMENTS DE LABORATOIRE A MAIN   33
      Tenailles   33
      Tenailles aux pinces en aiguilles   33
      Tenailles coupe-fer   33
      Sciseaux pour aluminium   33
      Sciseaux   33
      Clé anglaise   34
      Clé anglaise à bout ouvert   34
      Trousseau de clés anglaises   34
      Marteau   34
      Tourne-vis   34
      Perceuse   35
      Limeuse en métal   35
      Vis d'etablis   35
      Soudeuse   35
      Discussion   36
FOURNITURES DE BUREAU   37
      Appareil de papier collant   37
      Marqueur d'étiquettes   37
      Etiquettes   37
      Style à bille   37
      Crayon à cire   38
      Crayon-feûtre   38
      Règle   38
      Calculatrice   38
      Discussion   39
TABLEAU PERIODIQUE   40
      Tableau ou carte périodique   40
      Discussion   41
REACTIF   42
      Acide acétique   42
      Alcool   42
      Hydroxyde d'ammonium   42
      Hydroxyde de calcium   42
      Acide hydrochlorique   42
      Acide nitrique   42
      Hydroxyde de potasse   42
      Hydroxyde de sodium   42
      Acide sulfurique   42
      Désignation de reactif   43
      Discussion   44
MODELES MOLECULAIRES   45
SUPPORTS   46
      Support ou cadre   46
      Support de burette   46
      Support d'anneaux   46

SUPPORTS (suite)
    Trépied 47
    Pince de burette 47
    Support de pinces 47
    Pince 47
    Pince avec support 48
    Discussion 49
VERRES 50
    Canne de verre 50
    Tube capillaire 50
    Baguette de verre 50
    Raccord et douille 51
    Charnière 51
    Rodage normalisé 51
    Pince de Mohr 51
    Robinet d'arrêt en verre 52
    Condensateur 52
    Eprouvette-test 52
    Support d'éprouvette test 53
    Support à tubes d'essais 53
    Verre de montre 53
    Poignée de sécurité 53
    Couteau à canne de verre 54
    Discussion 55
RECIPIENTS 56
    Bécher 56
    Bouteille 56
    Falcon conique d'Erlenmeyer 56
    Seau 56
    Flacon de Florence 57
    Fiole 57
    Couvercle 57
    Bouchon d'arrêt en caoutchouc ou plastiue . . 57
    Bouchon d'arrêt en verre 57
    Bouchon 58
    Tire-bouchon 58
    Discussion 59
TUYAUTERIE 60
    Tuyauterie en caoutchouc 60
    Tuyauterie en plastique 60
    Pince de tuyau 60
    Discussion 61
INSTRUMENTS DE MESURE DE VOLUME LIQUIDE 62
    Pipette en seringue 62
    Pipette volumétrique 62
    Pipette graduée 62
    Ballon de pipette 63
    Pipette de transfert automatique 63
    Cylindre gradué 63
    Récipient gradué 64
    Bécher 64
    Burette graduée 64
    Pipette au zero automatique 65

EQUIPEMENT DE COMPRESSION ET DE MELANGE (suite)
    Homogénéiseur à emulsion   84
    Presse hydraulique   85
    Discussion   86
INSTRUMENTS DE MESURE   87
    Appareil de mesure du pH   87
    Electrode   87
    Solution tampon pH   87
    Papier indicateur   88
    Voltamètre Ohm   88
    Baromètre   88
    Calorimètre   89
    Centrifugeuse   89
    Tête de centrifugeuse   90
    Tube de centrifugeuse   90
    Porteurs   90
    Spectrophomètre   91
    Cellule de spectrophomètre   91
    Fluoromètre   92
    Débimètre   92
    Discussion   93
INSTRUMENTS DE SEPARATION   94
    Electrophorèse à colonne de gel   94
    Densitomètre d'analyse   94
    Dialyse   94
    Echantilloneur automatique   95
    Tamis   95
    Filtre de séparation   95
    Colonne de chromatographie   96
    Chromatographie à couche mince   96
    Cabinet de chromatographie   96
    Cartouche de Soxhlet   97
    Cartouche d'extraction   97
    Extracteur de Goldfisch   97
    Bouteille de Babcock   98
    Diviseur   98
    Centrifugeuse de Babcock   98
    Assiette en papier aluminium   99
    Dessicateur   99
    Etuve   99
    Etuve à vide 100
    Degré moyen d'humidite 100
    Séchage par évaporation de glace 100
    Flacon de Kjeldahl 101
    Unité de digestion de Kjeldahl 101
    Unité de distillation de Kjeldahl 101
    Indicateur 102
    Discussion 103
FILTRAGE 104
    Entonnoir 104
    Papier filtre 104
    Laine de verre 104
    Ballon filtre 105

FILTRAGE (suite)
    Entonnoir filtre de Buchner 105
    Creuset 105
    Support de creuset 106
    Epurateur 106
    Discussion 107
DISTILLATION 108
    Alambic 108
    Flacon de distillation 108
    Alambic à eau 108
    Déminéralisateur 109
    Evaporateur instantané 109
    Discussion 110
MESURE DE GAZ 111
    Appareil respiratoire 111
    Discussion 112
EQUIPEMENT POUR PETIT ANIMAL 113
    Cage 113
    Tube d'alimentation en eau 113
    Assiette 113
    Trousse de dissection 114
    Table d'opération pour animaux 114
    Discussion 115
EXAMEN MICROSCOPIQUE 116
    Microscope 116
    Diapositive 116
    Loupe 116
    Tissu de lentille 117
    Micromètre 117
    Cuve de comptage 117
    Flacon ou plat de coloration 118
    Unité de traitement des tissus 118
    Microtome 118
    Discussion 119
MICROBIOLOGIE 120
    Milieu de culture 120
    Bouteille de dilution 120
    Plat de Petri 120
    Boîte de stérilisation du plat de Petri . . . 121
    Compteur de la colonie de Québec 121
    Compteur 121
    Aiguille d'inoculation 122
    Tube de culture 122
    Discussion 123
LAVAGE 124
    Laveur de verres 124
    Laveur de pipettes 124
    Egoutteur 125
    Discussion 126

ARBEITSSICHERHEIT AUSSTATUNG   1
    "Erste Hilfe" Tasche   1
    Feuerlöscher   1
    Feuerdecke   1
    Dusche   1
    Augenspülwasser   2
    Telefon   2
    Warntafeln, Warnzeichen   2
            Vorsicht   2
            Sprengstoff   2
            Gefahr der biologischen Verseuchung . . .   2
            Gefahr der radioaktiven Verstrahlung . . .   3
            Setz die Schutzbrille auf   3
            Gefahr   3
            Leicht anzündbar   3
            Zerfressend   3
            Gift   3
            Lüftung nötig   3
            Unter hohem Druck   3
            Unter niedrigem Druck   3
            Kaustisch   3
            Berührung der Haut vermeiden   3
            Heiß   3
            Steril   3
            Im Kühlschrank aufbewahren   3
            Kaputt   3
        Schutzbrille   4
        Atemgerät   4
        Diskussion   5
LABORMÖBEL   6
        Abzug   6
        Laborschrank mit Arbeitsbrett   6
        Lagerschrank   6
        Schrank   7
        Schreibtisch   7
        Tisch   7
        Aktenschrank   7
        Diskussion   8
TRAGBARE LABORMÖBEL   9
        Laborwagen   9
        Laborstuhl   9
        Leiter   9
        Diskussion   10
ARMATUR   11
        Hahn   11
        Warm und kalt Wasserhahn   11
        Spühlbecken   11
        Absperrhahn   12
        Kaltes Wasser   12
        Heißes Wasser   12
        Destilliertes Wasser   12
        Dampf   12
        Luft   12
        Gas   13
ARMATUR (Fortzetzung)

Sauerstoff (O₂)   13
Stickstoff (N₂)   13
Vakuum 13
Nadelventil   13
Siphon 13
Diskussion 14
ELEKTRISCHE AUSSTATTUNG   15
Anschluß 15
Anschlußkasten 15
Variabel Transformator ("Variac")   15
Batterie; Aufladenbar und hichtaufladenbar   . .   16
Ladegerät   16
Diskussion 17
FLASCHENGAS   18
Gasflasche oder Gasbehälter   18
Druckregulator   18
Klammer der Gasflasche   18
Diskussion   19
VAKUUM- UND LUFTPUMPEN 20
Vakuum-und Druckpumpe   20
Bewegung des Pumpenrotors   20
Vakuumpumpe (Drehkolbenölpumpe)   20
Handpumpe   21
Stützbare Handpumpe   21
Aspirator (Wasserpumpe)   21
Diskussion   22
LAGERUNG IN NIEDRIGEN TEMPERATUREN 23
Kühlschrank-Tiefkühltruhe   23
Trockeneisbehälter (Kohlendioxidschneebehälter)   23
Trockeneiskasten (zur Lagerung und Transport) .   24
Dewargefäß   24
Diskussion   25
LABOR HANDAUSRÜCTUNG 26
Spachtel   26
Schaufel   26
Zange   26
Rührer   26
Notizheft   26
Folie   27
Tüten   27
Abdampfschale   27
Schürze   27
Asbesthandschuhe   28
Gummihanschuhe   28
Handschuhe zum einmaligen Gebrauch   28
Reinigungsflasche   28
Tropfflasche   29
Gummibirne   29
Bürste   29
Putzmittel   29
Papiertücher   29
Wanduhr   30
Intervalstoppuhr   30
Zeitschalter   30
LABOR HANDAUSRÜCTUNG (Fortzetzung)

Stoppuhr 31
Diskussion 32
LABORWRKZEUG 33
Zange 33
Nadelförmige Zange 33
Drahtschneidende Zange 33
Metalschneidende Schere 33
Schere 33
Schraubenschlüssel (einstellbar) 34
Schraubenschlüssel 34
Flachschlüsselgarnitur 34
Hammer 34
Schraubenzieher 34
Bohrenspitze 35
Metalfeile 35
Aufspannblock 35
Lötpistole 35
Diskussion 36
BüROWERKZEUG 37
Klebebandverteiler 37
Beschriftungsgerät 37
Aufklebezettel 37
Kugelschreiber 37
Fettstift 38
Filzstifft 38
Linial 38
Rechner 38
Diskussion 39
PERIODENSYSTEM 40
Periodensystem oder Tabelle 40
Diskussion 41
REKTIONSMITTEL 42
Essigsäure 42
Alkohol 42
Ammoniumhydroxid 42
Calciumhydroxid 42
Salzsäure 42
Salpetersäure 42
Kaliumhydroxid 42
Natriumhydroxid 42
Schwefelsäure 42
Reagenspapier 43
Diskussion 44
MOLEKüLMODELLE 45
Hilfsmittel 46
Stativ 46
Bürettengestell 46
Retortengestell 46
Dreibein 47
Bürettenklammer 47
Klammerhalter 47
Klammer 47
Klammet mit Halter 48
Diskussion 49
GLASS 50

Glassrohr 50
Haarrörchen 50
Glasstab 50
Kugelgelenk 51
O-Ringgelenk 51
Standard Verjüngtes Glasschliff 51
Quetschhahn 51
Glasshahn 52
Kühler 52
Reagenzglass 52
Reagenzglasshalter 53
Reagenzglassgestell 53
Uhrglass 53
Handschutz 53
Glassrohrfräser 54
Diskussion 55
BEHÄLTER UND VERSCHLüSSE 56
Becher 56
Flasche 56
Erlenmeyerkolben 56
Kübel oder Eimer 56
Flachbodenkolben 57
Ampulle 57
Kappe 57
Stöpsel; Gummi- und Plastikstöpsel 57
Glasstöpsel 57
Korken 58
Korkbohrer 58
Diskussion 59
ROHRLEITUNG 60
Gummischlauch 60
Plastikschlauch 60
Schlauchhahn 60
Diskussion 61
FLüSSIGKEITSMASSAUSRüSTUNG 62
Spritzpipette 62
Vollpipette 62
Messpipette 62
Pipettenbirne 63
Automatische Vollpipette 63
Messzylinder 63
Messglass 64
Becher 64
Messbürette 64
Automatische Nullpipette 65
Abspeerhahn 65
Abspeerhahnschmiere 65
Messkolben 66
Pumpe 66
Diskussion 67
GEWICHT UND ZUSAMMENSETZUNG DER PROBE 68
Federwaage 68
Hebelwaage 68
Analytische Waage 68
GEWICHT UND ZUSAMMENSETZUNG (Fortzetzung)

Automatische Waage   69
Gewichtssatz   69
Wichtewaage   69
Hydrometer   70
Pyknometer   70
Viscozimeter   70
Polarimeter   71
Refraktometer   71
Handrefrektometer   71
Diskussion   72
ERHITZEN   73
Ofen   73
Heizplatte   73
Brutapparat   73
Heizpilz   74
Tauchsieder   74
Heizband   74
Muffelofen   75
Schmelztiegel   75
Wasserbad   75
Thermometer   76
Thermometer mit Zifferblatt   76
Elektronisches Thermometer   76
Sterilisator   77
Sterilisator der Instrumente   77
Brenner   77
Spirituslampe   78
Gasanzünder   78
Streichhölzer   78
Schmelzpunktbestimmungsapparat   79
Schmelzpunktrörchen   79
Diskussion   80
MISCH- UND PRESSAUSRüSTUNG   81
Mörser und Mörserkeule   81
Rührer   81
Magnetischer Rührer   81
Grosse Schüttelvorrichtung   82
Schüttelvorrichtung   82
Mixer   82
Handschleifmaschine   83
Mühle   83
Gewbwschleifmaschine   83
Homogenisierapparat   84
Emmulsionshomogenisiersapparat   84
Hydraulische Presse   85
Diskussion   86
MEBAUSRüSTUNG   87
pH-Meter   87
Elektrode   87
pH-Pufferlösung   87
Lackmuspapier   88
Volt-Ohm-Amperameter   88
Barometer   88
Bombenkalorimeter   89
MEBAUSRüSTUNG (Fortzetzung)

Zentrifuge 89
Zentrifugenkopf (Rotor) 90
Zentrifugenglass 90
Träger 90
Spektralphotometer 91
Spektralzelle 91
Fluorometer 92
Durchflußmesser 92
Diskussion 93
TRENNAUSRüSTUNG 94
Gel-Säuleelektrophorese 94
Abtastdenitometer 94
Dialysator 94
Automatischer Probennehmer 95
Sieb 95
Scheidetrichter 95
Trennsäule 96
Dünnschichtchromatographie 96
ChromatographieKammer 96
Soxhlet-Extraktor 97
Extraktionshülse 97
Goldfisch-Extraktor 97
Babcock-Flasche 98
Stechzirkel 98
Babcock-Zentrifuge 98
Behälter aus Aluminiumfolie 99
Exsikkator 99
Trockner 99
Vakuumofen 100
Feuchtigkeitswaage 100
Gefriertrocknung 100
Kjeldahl-Kolben 101
Kjeldahl-Digestioneinheit 101
Kjeldahl-Destilliereinheit 101
Registrierapparat 102
Diskussion 103
FILTRIEREN 104
Trichter 104
Filterpapier 104
Glasswolle 104
Absaugeflasche 105
Büchner-Filtertrichter 105
Glühtiegel 105
Tiegalhalter 106
Staubprobansammler 106
Diskussion 107
DESTILLATION 108
Retorte 108
Destillierkolben 108
Wasserdestillierappart 108
Entsaltzungsvorrichtung 109
Stoßverdampfer 109
Diskussion 110

GASMESSUNG 111

    Respiratometer 111
    Diskussion 112
ASRÜSTUNG ZU VERSUCHEN MIT KLEINEN TIEREN 113
    Käfig 113
    Tränkrohr 113
    Fütterschale 113
    Präpariersatz 114
    Operationstisch für Tiere 114
    Diskussion 115
MIKROSKOPISCHE UNTERSUCHUNG 116
    Mikroskop 116
    Schieber 116
    Lupe 116
    Linsentuch 117
    Mikrometer 117
    Zählkammer 117
    Färbetopf 118
    Gewebebearbeitungseinheit 118
    Mikrotom 118
    Diskussion 119
MIKROBIOLOGIE 120
    Kulturböden 120
    Verdünnungsflasche 120
    Petrischale 120
    Sterilisationskasten für Petrischale 121
    Quebec-Kolonienzähler 121
    Zähler 121
    Impfungsnadel oder Schlinge 122
    Kulturrörchen 122
    Diskussion 123
WASCHEN 124
    Waschvorrichtung für Glass 124
    Waschvorrichtung für Pipette 124
    Trockner für Glass 125
    Diskussion 126

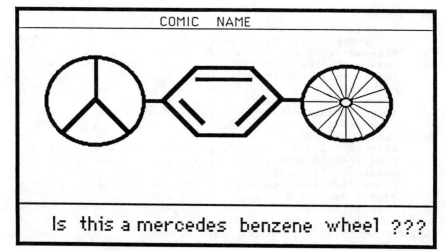

COMIC   NAME

Is this a mercedes benzene wheel ???

WYPOSAŻENIE ZWIĄZANE Z BEZPIECZEŃSTWEM PRACY   1
    Zestaw pierwszej pomocy medycznej   1
    Gaśnica   1
    Koc przeciwpożarowy   1
    Prysznic   1
    Woda do płukania oczu   2
    Telefon   2
    Napisy i znaki ostrzegające   2
        ostrożnie   2
        materiał wybuchowy   2
        niebezpieczeństwo skażenia biologicznego   2
        zagrożenie promieniowaniem   3
        załóż okulary ochronne   3
        niebezpieczeństwo   3
        zagrożenie pożarowe   3
        substancja korozyjna   3
        trucizna   3
        konieczna odpowiednia wentylacja   3
        pod wysokim ciśnieniem   3
        pod niskim ciśnieniem   3
        żrąca zasada   3
        unikać kontaktu ze skórą   3
        wysoka temperatura   3
        steryine   3
        przechowywać w lodówce   3
        uszkodzone   3
    Okulary ochronne   4
    Wziewnik   4
    Dyskusja   5
MEBLE LABORATORYJNE   6
    Wyciąg   6
    Szafka laboratoryjna z blatem roboczym . . .   6
    Wysoka szafka laboratoryjna   6
    Wysoka szafka laboratoryjna z przesuwanymi
        drzwiami   7
    Biurko   7
    Stół   7
    Szafka z szufladami do przechowywania
        dokumentów   7
    Dyskusja   8
PRZENOŚNE MEBLE LABORATORYJNE   9
    Ruchomy stolik laboratoryjny   9
    Krzesło laboratoryjne   9
    Drabinka   9
    Dyskusja   10
WYPOSAŻENIE   11
    Kran wodociągowy   11
    Kran wodociągowy z zimną i ciepłą wodą . . .   11
    Zlew   11
    Kurek   12
    Woda zimna   12
    Woda gorąca   12
    Woda destylowana   12
    Para wodna   12
    Powietrze   12
    Gaz   13

WYPOSAŻENIE(c.d.)
  Tien(O_2)   13
  Azot(N_2)   13
  Próżnia   13
  Zawór iglicowy   13
  Rurka przelewowa, syfon   13
  Dyskusja   14
URZĄDZENIA ELEKTRYCZNE   15
  Gniazdko elektryczne   15
  Rozdzielacz, kontakt przenośny   15
  Transformator   15
  Bateria   16
  Urżądzenie do ładowania baterii   16
  Dyskusja   17
GAZ W BUTLACH   18
  Butla gazowa   18
  Regulator ciśnienia   18
  Uchwyt do mocowania butli gazowej   18
  Dyskusja   19
POMPY PRÓŻNIOWE I POWIETRZNE   20
  Pompa próżniowo-ciśnieniowa   20
  Ruch wirnika pompy   20
  Pompa próżniowa   20
  Pompka ręczna   21
  Pompka ręczna (opierana o podłogę)   21
  Pompka wodna   21
  Dyskusja   22
PRZECHOWYWANIE W NISKICH TEMPERATURACH   23
  Lodówka z zamrażalnikiem   23
  Pojemnik do przechowywania suchego lodu .   23
  Pudło do transportu i przechowywania
    suchego lodu   24
  Termos   24
  Dyskusja   25
PODRĘCZNY SPRZĘT LABORATORYJNY   26
  Szpachelka   26
  Szufelka   26
  Szczypce   26
  Mieszadło   26
  Notatnik   26
  Folia   27
  Torebka   27
  Parownica   27
  Fartuch ochronny   27
  Rękawice azbestowe   28
  Rękawice gumowe   28
  Rękawice jednorazowe   28
  Tryskawka   28
  Kroplomierz   29
  Gruszka gumowa   29
  Szczotka   29
  Środek zmywający   29
  Bibułka do czyszczenia   29
  Zegar ścienny   30
  Czasomierz, minutnik   30
  Wyłącznik zegarowy   30
  Stoper   31
  Dyskusja   32

PODRĘCZNE NARZĘDZIA LABORATORYJNE   33
    Szczypce   33
    Szczypce do wyginania i cięcia drutu   33
    Szczypce do cięcia drutu   33
    Nożyce do cięcia blachy   33
    Nożyczki   33
    Klucz francuski   34
    Klucz płaski do nakrętek   34
    Zestaw kluczy wpustowych sześciokątnych .   34
    Młotek   34
    Śróbokręt   34
    Wiertło   35
    Pilnik   35
    Imadło   35
    Lutownica   35
    Dyskusja   36
PRZYBORY BIUROWE   37
    Przyrząd do rozwijania i cięcia taśmy
        klejącej   37
    Przyrząd do tłoczenia napisow na
        plastikowych nalepkach   37
    Nalepka   37
    Długopis   37
    Ołówek woskowy, dermatograf   38
    Pisak, flamaster   38
    Linijka   38
    Kalkulator   38
    Dyskusja   39
TABELA UKLADU OKRESOWEGO   40
    Tabela układu okresowego pierwiastków . .   40
    Dyskusja   41
ODCZYNNIKI CHEMICZNE   42
    Kwas octowy   42
    Alkohol   42
    Wodorotlenek amonowy   42
    Wodorotlenek wapniowy   42
    Kwas solny   42
    Kwas azotowy   42
    Wodorotlenek potasowy   42
    Wodorotlenek sodowy   42
    Kwas siarkowy   42
    Etykiety na odczynnikach   43
    Dyskusja   44
MODELE MOLEKULARNE   45
STATYWY   46
    Statyw ramowy   46
    Statyw do biuret   46
    Statyw laboratoryjny   46
    Trójnóg   47
    Uchwyt, łapa do biuret   47
    Łącznik   47
    Łapa statywu   47
    Łapa statywu z łącznikiem   48
    Dyskusja   49

SZKŁO LABORATORYJNE 50
    Rurka szklana 50
    Rurka kapilarna 50
    Bagietka szklana 50
    Połączenie na szlif kulisty 51
    Łącznik pierścieniowy 51
    Połączenie na standardowy szlif stożkowy . 51
    Klamra zaciskowa 51
    Kurek szklany 52
    Chłodnica 52
    Probówka 52
    Uchwyt do probówek 53
    Statyw do probówek 53
    Szkiełko zegarkowe 53
    Ochraniacz dłoni 53
    Urządzenie do cięcia szklanych rurek 54
    Dyskusja 55
NACZYNIA I ICH ZAMKNIĘCIA 56
    Zlewka 56
    Butelka 56
    Kolba stożkowa 56
    Wiadro 56
    Kolba płaskodenna 57
    Fiolka 57
    Wieczko, nakrętka 57
    Korek 57
    Korek szklany 57
    Korek z kory drzewa 58
    Korkobór 58
    Dyskusja 59
RURKI 60
    Rurka gumowa 60
    Rurka plastikowa 60
    Zaciskacz do rurka 60
    Dyskusja 61
SPRZĘT DO MIERZENIA OBJĘTOŚCI CIECZY 62
    Pipeta strzykawkowa 62
    Pipeta miarowa 62
    Pipeta wielomiarowa 62
    Pipeta napełniana gruszką gumową 63
    Dozownik 63
    Cylinder miarowy 63
    Menzurka 64
    Zlewka 64
    Biureta wielomiarowa 64
    Pipeta automatyczna jednomiarowa 65
    Kurek 65
    Smar do uszczelniania kurków 65
    Kolba miarowa 66
    Pompka 66
    Dyskusja 67

CIĘŻAR I SKŁAD PRÓBKI   68
   Waga sprężynowa   68
   Waga trójramienna z przesuwnymi
      odważnikami   68
   Waga analityczna   68
   Waga automatyczna jednoszalkowa   69
   Zestaw odważników   69
   Waga do określania ciężaru właściwego . . .   69
   Areometr   70
   Piknometr   70
   Wiskozymetr   70
   Polarymetr   71
   Refraktometr   71
   Refraktometr ręczny   71
   Dyskusja   72
OGRZEWANIE   73
   Suszarka   73
   Podgrzewacz   73
   Cieplarka   73
   Płaszcz grzejny   74
   Grzejnik zanurzeniowy   74
   Taśma grzejna   74
   Piec muflowy   75
   Tygiel   75
   Łaźnia wodna   75
   Termometr   76
   Termometr tarczowy   76
   Termometr elektroniczny   76
   Autoklaw   77
   Sterylizator do narzędzi   77
   Palnik   77
   Lampka spirytusowa   78
   Zapalacz do gazu   78
   Zapałki   78
   Przyrząd do oznaczania temperatury
      topnienia   79
   Probówka do oznaczania temperatury
      topnienia   79
   Dyskusja   80
SPRZĘT DO MIESZANIA I ZGNIATANIA   81
   Moździerz i pistol   81
   Mieszadło   81
   Mieszadło magnetyczne   81
   Duża wstrząsarka   82
   Wstrząsarka.   82
   Mikser   82
   Ręczna maszynka do mielenia   83
   Młynek   83
   Homogenizator szklany do miękkich tkanek .   83
   Homogenizator   84
   Homogenizator emulsyjny   84
   Prasa hydrauliczna   85
   Dyskusja   86

DESTYLACJA 108
    Aparat destylacyjny 108
    Kolba destylacyjna 108
    Destylarka 108
    Demineralizator 109
    Wyparka 109
    Dyskusja 110
BADANIE PRZEMIAN GAZOWYCH 111
    Aparat Warburga 111
    Dyskusja 112
SPRZĘT DO BADAŃ Z UŻYCIEM MAŁYCH ZWIERZĄT 113
    Klatka dla zwierząt 113
    Poidełko 113
    Karmidełko 113
    Zestaw sekcyjny 114
    Stół operacyjny dla zwierząt 114
    Dyskusja 115
BADANIA MIKROSKOPOWE 116
    Mikroskop 116
    Szkiełko podstawowe 116
    Lupa 116
    Bibułka do czyszczenia soczewek 117
    Mikrometr 117
    Komora do liczenia krwinek i drobnoustrojów 117
    Naczynie do barwienia 118
    Aparat do przygotowywania preparatów
        histologicznych 118
    Mikrotom 118
    Dyskusja 119
MIKROBIOLOGIA 120
    Podłoża 120
    Butelka do rozcieńczeń 120
    Płytka Petriego 120
    Pojemnik do sterylizacji płytek Petriego . . 121
    Licznik kolonii 121
    Licznik 121
    Eza 122
    Probówka bakteriologiczna 122
    Dyskusja 123
ZMYWANIE 124
    Zmywarka do szkła laboratoryjnego 124
    Płuczka do pipet 124
    Wieszak do suszenia szkła laboratoryjnego 125
    Dyskusja 126

EQUIPO DE SEGURIDAD   1
    Botiquín   1
    Extintor de incendios   1
    Manta contra incendios   1
    Ducha   1
    Lavaojos   2
    Teléfono   2
    Etiquetas y señales   2
        Precaución   2
        Explosivo   2
        Riesgo Biológico   2
        Riesgo Radiactivo   3
        Use Gafas   3
        Peligro   3
        Inflamable   3
        Corrosivo   3
        Veneno   3
        Usar en la vitrina de gases   3
        A presion elevada   3
        A presion reducida   3
        Cáustico   3
        Evite el contacto con la piel   3
        Caliente   3
        Estéril   3
        Mantener en refrigeracion   3
        No Funciona   3
    Gafas protectoras   4
    Mascarilla   4
    Discusión   5
MOBILIARIO DE LABORATORIO   6
    Vitrina de gases   6
    Módulo de sustentacion   6
    Armario   6
    Estantería   7
    Mesa de despacho   7
    Mesa   7
    Archivador   7
    Discusión   8
MATERIAL PORTATIL DE LABORATORIO   9
    Carro de laboratorio   9
    Silla o taburete de laboratorio   9
    Escalera   9
    Discusión   10
INSTALACION DE FONTANERIA   11
    Grifo   11
    Grifo combinado   11
    Fregadero   11
    Grifo   12
    Agua fría   12
    Agua caliente   12
    Agua destilada   12
    Vapor   12
    Aire   12

INSTALACION DE FONTANERIA
    Gas 13
    Oxígeno 13
    Nitrógeno 13
    Vacío 13
    Válvula de aguja 13
    Sifón 13
    Discusión 14
INSTALACION ELECTRICA 15
    Enchufe 15
    Caja de conexiones eléctricas 15
    Transformador 15
    Bateria 16
    Cargador de baterías 16
    Discusión 17
GASES LICUADOS 18
    Cilindro de gas 18
    Regulador de presión 18
    Pinza de sujeción para cilindros de gas . . . 18
    Discusión 19
BOMBAS DE VACIO Y COMPRESORES 20
    Bomba de vacío y compresor 20
    Bomba Peristáltica 20
    Bomba de vacío 20
    Bomba manual 21
    Bomba de vacío de Agua 21
    Discusión 22
ALMACENAMIENTO FRIGORIFICO 23
    Frigorífico-Congelador 23
    Arcón para hielo seco 23
    Recipiente para hielo seco 24
    Termo 24
    Discusión 25
EQUIPO MANUAL DE LABORATORIO 26
    Espátula 26
    Cucharón 26
    Pinzas para crisoles 26
    Cuchara de agitación 26
    Libro de notas 26
    Hoja de aluminio 27
    Bolsa 27
    Cápsula 27
    Delantal 27
    Guantes de amianto 28
    Guantes de goma 28
    Guantes desechables 28
    Frasco lavador 28
    Frasco cuentagotas 29
    Pera de goma 29
    Escobillón 29
    Limpiador 29
    Pañuelo 29
    Reloj de pared 30
    Reloj avisador 30
    Programador 30

EQUIPO MANUAL DE LABORATORIO
    Cronómetro   31
    Discusión   32
HERRAMIENTAS DE LABORATORIO   33
    Alicates   33
    Alicates de punta fina   33
    Tenazas   33
    Tijeras   33
    Llave Ajustable   34
    Llave de dos bocas   34
    Llaves para tuercas hexagonales   34
    Martillo   34
    Destornillador   34
    Broca   35
    Lima   35
    Tornillo de banco   35
    Soldador   35
    Discusión   36
MATERIAL DE OFICINA   37
    Cinta Adhesiva   37
    Etiquetador   37
    Etiqueta   37
    Bolígrafo   37
    Lapiz de cera   38
    Rotulador   38
    Regla   38
    Calculadora   38
    Discusión   39
TABLA PERIODICA   40
    Tabla Periódica   40
    Discusión   41
REACTIVOS   42
    Acido Acético   42
    Alcohol   42
    Hidróxido Amónico   42
    Hidróxido Cálcico   42
    Acido Clorhídrico   42
    Acido Nítrico   42
    Hidróxido Potásico   42
    Hidroxido Sódico   42
    Acido Sulfúrico   42
    Etiquetado de Reactivos   43
    Discusión   44
MODELOS MOLECULARES   45
SOPORTES   46
    Soporte o Emparrillado   46
    Soporte para buretas   46
    Soporte de aro   46
    Trípode   47
    Pinza para buretas   47
    Nuez   47
    Pinza   47
    Pinza con nuez   48
    Discusión   49

MATERIAL DE VIDRIO   50
    Tubo de vidrio   50
    Tubo Capilar   50
    Varilla de vidrio   50
    Union de rótula   51
    Union plana   51
    Union cónica esmerilada   51
    Pinza para esmerilados de rótula   51
    Llave de paso de vidrio   52
    Condensador   52
    Tubo de ensayo   52
    Pinzas para tubos de ensayo   53
    Gradilla   53
    Vidrio de reloj   53
    Mango de seguridad   53
    Cortadora de tubos de vidrio   54
    Discusión   55
ENVASES Y RECIPIENTES   56
    Vaso de precipitado   56
    Botella   56
    Matraz erlenmeyer   56
    Cubo con asa   56
    Matraz de fondo plano   57
    Vial   57
    Tapa   57
    Tapon   57
    Tapon de vidrio   57
    Corcho   58
    Taladratapones   58
    Discusión   59
TUBOS   60
    Tubo de goma   60
    Tubo de plástico   60
    Pinza de mohr   60
    Discusión   61
EQUIPO PARA VOLUMETRIA   62
    Jeringuilla micrométrica   62
    Pipeta aforada   62
    Pipeta graduada   62
    Pera   63
    Pipeta automática   63
    Probeta   63
    Copa graduada   64
    Vaso de precipitado   64
    Bureta   64
    Pipeta automática   65
    Llave de vidrio   65
    Grasa para buretas   65
    Matraz volumétrico   66
    Bomba   66
    Discusión   67
EQUIPOS PARA PESADA Y MEDIDA   68
    Báscula   68
    Granatario   68
    Balanza analítica   68

EQUIPOS PARA PESADA Y MEDIDA
    Balanza digital 69
    Juego de pesas 69
    Balanza para densidades 69
    Areómetro 70
    Picnómetro 70
    Viscosímetro 70
    Polarímetro 71
    Refractómetro 71
    Refractómetro manual 71
    Discusión 72
CALENTADORES 73
    Horno 73
    Placa calentadora 73
    Estufa 73
    Manta calefactora 74
    Termostato 74
    Cinta calefactora 74
    Horno de mufla 75
    Crisol 75
    Baño maría 75
    Termómetro 76
    Termómetro industrial 76
    Termómetro electrónico 76
    Autoclave 77
    Esterilizador de instrumentos 77
    Mechero 77
    Lámpara de alcohol 78
    Encendedor 78
    Fósforos de seguridad 78
    Baño para punto de fusión 79
    Tubo para punto de fusión 79
    Discusión 80
EQUIPO PARA MEZCLA Y PRENSADO 81
    Mortero y mano 81
    Agitador 81
    Agitador magnético 81
    Agitador-Vibrador 82
    Agitador-Vibrador 82
    Batidora 82
    Picadora manual 83
    Molino 83
    Homogeneizador de vidrio 83
    Homogeneizador 84
    Emulsificador 84
    Prensa hidráulica 85
    Discusión 86
EQUIPO DE MEDIDA 87
    pH-Metro 87
    Electrodo 87
    Solución tampón 87
    Papel indicador 88
    Voltímetro-Amperímetro 88
    Barómetro 88

EQUIPO DE MEDIDA
   Bomba calorimétrica 89
   Centrífuga 89
   Cabezal de centrífuga 90
   Tubo de centrífuga 90
   Cabezal múltiple (centrífuga) 90
   Espectrofotómetro 91
   Cubeta para espectrofotometría 91
   Espectrofluorímetro 92
   Flotámetro 92
   Discusión 93
EQUIPO DE SEPARACION 94
   Equipo para electroforesis en gel 94
   Densitómetro 94
   Equipo de diálisis 94
   Equipo de muestreo automático 95
   Criba 95
   Embudo de separación 95
   Columna cromatográfica 96
   Cromatografía en capa fina 96
   Cámara cromatográfica 96
   Extractor de Soxhlet 97
   Cartucho de extracción 97
   Extractor de Goldfisch 97
   Matraz de Babcock 98
   Compás de puntas 98
   Centrífuga de Babcock 98
   Platillo de hoja de aluminio 99
   Desecador 99
   Estufa de desecación 99
   Estufa de vacío 100
   Balanza para humedad 100
   Liofilizador 100
   Matraz de Kjeldahl 101
   Equipo de digestión de Kjeldahl 101
   Equipo de destilación de Kjeldahl 101
   Registrador 102
   Discusión 103
EQUIPO DE FILTRACION 104
   Embudo 104
   Papel de filtro 104
   Lana de vidrio 104
   Matraz Kitasato 105
   Embudo de Buchner 105
   Crisol de Gooch 105
   Conos de goma para embudos Buchner 106
   Frasco lavador de gases 106
   Discusión 107
EQUIPO DE DESTILACION 108
   Aparato para destilación 108
   Matraz de destilación 108
   Destilador de agua 108
   Desmineralizador 109
   Evaporador 109
   Discusión 110

Extinct

Is forever

EMNİYET ALET VE EKİPMANLARI
İlk yardim kutusu 1
Yangın söndürücü 1
Yangın örtüsü 1
Duş için çeşme 1
Göz yıkamak için çeşme 2
Telefon 2
Tehlike etiket ve işaretleri 2
    tedbirli ol (caution) 2
    patlayıcı (explosive) 2
    biyolojik tehlike (biohazard) 2
    radyasyon tehlikesi (radiation hazard) 3
    koruyucu gözlüğünü tak (wear your goggles) 3
    tehlike (danger) 3
    alevlenebilir (flammable) 3
    aşındırıcı (corrosive) 3
    zehir (poison) 3
    çeker ocağı kullan 3
    yüksek basınç 3
    alçak basınç 3
    kostik (caustic) 3
    deri temasından sakın 3
    sıcak (hot) 3
    steril 3
    buzdolabında muhafaza et 3
    bozuk 3
Emniyet gözlüğü 4
Teneffüs cihazı 4
Tartışma 5
LABORATUVAR DONANIMI 6
Çeker ocak 6
Çalışma ünitesi 6
Malzeme dolabı 7
Ofis tipi çalışma masası 7
Masa 7
Büro tipi dosya dolabı 7
Tartışma 8
TAŞINABİLİR LABORATUVAR DONANIMI 9
Laboratuvar tipi el arabası 9
Laboratuvar tipi sandalye 9
Merdiven 9
Tartışma 10
SABİT LABORATUVAR DONANIMI 10
Musluk 11
Sıcak soğuk karıştırma musluğu 11
Lavoba 11
Dişli musluk (sızmayi önleyen) 12
Soğuk su 12
Sıcak su 12
Distile su 12

SABİT DONANIM (devam)
Buhar 12
Hava 12
Havagazı 13
Oksijen (O_2) 13
Azot (N_2) 13
Vakum 13
Vana 13
Sifon 13
Tartışma 14
ELEKTRİK DONANIMI **15**
Priz 15
Çoklu priz 15
Transformotor 15
Pil (şarz olabilir veya olamaz tip) 16
Pil şarjer (şarz aleti) 16
Tartışma 17
GAZ TÜPLERİ **18**
Tüp 18
Basınç ayarlayıcısı 18
Gaz tüpü emniyet kemeri 18
Tartışma 19
VAKUM ve HAVA POMPALARI **20**
Vakum ve basınç pompası 20
Pompa hareket vanası 20
Vakum pompası 20
El pompası 21
Ayak pompası 21
Su trompu 21
Tartışma 22
SOĞUK DEPOLAMA **23**
Buzdolabi ve derin dondurucu 23
Kuru buz (CO_2) dolabı 23
Kuru buz kabı 24
Termus 24
Tartışma 25
LABORATUVAR EL ARAÇ ve GEREÇLERİ **26**
Spatula 26
Kepce 26
Maşa 26
Kazıyıcı (spatula) 26
Not defteri 26
Aliminyum foil 27
Naylon torba 27
Porselen kap 27
Önlük 27
Asbest eldiven 28
Bulaşık eldiveni 28
Kullanılıp atılabilen eldiven 28
Plastik su şişesi 28

LABORATUVAR EL ARAÇ ve GEREÇLERİ (devam)
Damlalık 29
Lastik bulb 29
Fırça 29
Temizlik malzemesi (sabun vb) 29
Kağıt mendil 29
Duvar saati 30
Laboratuvar tipi saat 30
Zamanlayıcı 30
Kronometre 31
Tartışma 32
LABORATUVAR EL ALETLERİ **33**
Pense 33
Karga burnu 33
Tel keser (pense) 33
Teneke makası 33
Makas 33
Kurbağacık 34
Anahtar takımı 34
Tornavida tipi anahtar takımı 34
Çekiç 34
Tornavida 34
Matkap (kalem) seti 35
Eğe 35
Mengene 35
Elektrikli havya 35
Tartışma 36
BÜRO MALZEMESİ **37**
Seloteyp (bant) 37
Etiket yazıcısı 37
Etiket 37
Kalın uçlu kalem 37
Cama yazar kalem 38
Cetvel 38
Hesap makinası 38
Tartışma 39
PERİYODİK CETVEL **40**
Periyodik cetvel 40
Tartışma 41
KİMYASAL MADDELER **42**
Asetik asit 42
Alkol 42
Amonyum hidroksit 42
Kalsiyum hidroksit 42
Hidroklorik asit 42
Nitrik asit 42
Potasyum hidroksit 42
Sodyum hidroksit 42
Sülfürik asit 42
Kimyasal madde etiketi 43

KİMYASAL MADDELER   (devam)
Tartışma 44
MOLEKÜL MODELLERİ **45**
DESTEK veya DAYAMA EKİPMANLARI **46**
Dayama çerçevesi 46
Büret ayaklığı 46
Bilezikli dayama ayağı 46
Üç ayak (lı ızgara) 47
Büret kıskacı 47
Kıskaç tutucu 47
Kıskaç 47
Kıskaçlı büret tutucu 48
Tartışma 49
CAM MALZEME **50**
Cam boru 50
Kapillar cam boru 50
Cam çubuk 50
Top ve yuva şeklinde ek (cam malzelerde) 51
Contalı ekleme 51
Standart traşlı cam ekleme 51
Maşalı kıskaç 51
Cam vana 52
Soğutucu 52
Deney tüpü 52
Deney tüpü maşası 53
Tüp sehpası 53
Saat camı 53
Bir çeşit elcek (cam çubuk için) 53
Cam boru kesicisi 54
Tartışma 55
KAP TİPİ MALZEMELER **56**
Beherglas 56
Cam şişe 56
Erlenmayer 56
Kova 56
Cam balon 57
Ufak şişe 57
Kapak 57
Lastik tıpa 57
Cam tıpa 57
Mantar tıpa 58
Mantar delici 58
Tartışma 59
BORU (HORTUM) TAKIMI **60**
Lastik hortum 60
Plastik hortum 60
Lastik boru kıskacı 60
Tartışma 61
SIVI HACİM ÖLÇME MALZEMELERİ **62**
Şırınga tipi pipet 62
Hacim pipeti 62

SIVI HACİM ÖLÇME MALZEMELERİ   (devam)
Dereceli pipet 62
Pipet emicisi (zararli sıvılar için) 63
Otomatik pipet 63
Ölçü silindiri 63
Kaba dereceli silindir 64
Dereceli beherglas 64
Büret 64
Otomatik büret 65
Cam musluk 65
Desikatör yağı (gres) 65
Hacimli balon 66
Pompa 66
Tartışma 67
TARTIM ve ÖRNEK KOMPOZİSYONU 68
Terazi 68
Üç kollu terazi 68
Hassas terazi 68
Üstten kefeli hassas terazi 69
Ağırlık seti 69
Özgül ağırlık terazisi 69
Hidrometre 70
Piknometre 70
Viskozimetre 70
Polarimetre 71
Refraktometre 71
El refraktometresi 71
Tartışma 72
ISITMA CİHAZLARI 73
Kurutma fırını 73
Hot plate 73
Inkübatör 73
Isı mantosu 74
Su ısıtıcısı 74
Isı kuşağı 74
Yakma fırını 75
Kül kabı 75
Su banyosu 75
Termometre 76
Göstergeli termometre 76
Elektronik termometre 76
Otoklav 77
Kaynatıcı 77
Bunzen bak 77
Alkol veya gazyağı lambası 78
Çakmak 78
Kiprit 78
Erime noktası tayini için banyo 79
Erime noktası tayin tüpü. 79
Tartışma 80

EZME, PARÇALAMA ve KARIŞTIRMA ALETLERİ **81**
Havan ve tokmak 81
Karıştırıcı 81
Magnetik karıştırıcı 81
Çalkalayıcı 82
Çalkalayıcı 82
Blender 82
El değirmeni 83
Değirmen 83
Doku öğütücüsü 83
Homojenizatör 84
Emülsiyon homojenizatörü. 84
Hidrolik pres 85
Tartışma 86
ÖLÇÜ ALET ve CİHAZLARI **87**
pH metre 87
Elektrot 87
Buffer çözeltisi 87
Turnusol kağıdı 88
Volt metre 88
Barometre 88
Bomb kalorimetresi 89
Santrifüj 89
Santrifüj başı 90
Santrifüj tüpü 90
Taşiyıcı 90
Spektrofotometre 91
Spektrofotometre tüpü 91
Florometre 92
Flowmetre 92
Tartışma 93
AYIRMA ALET VE CİHAZLARI **94**
Jel elektroforezi 94
Scanning densitometre 94
Diyaliz cihazı 94
Otomatik örnek alıcı 95
Elek 95
Ayirma hunisi 95
Kromotografi kolonu 96
İnce tabaka kromotografisi 96
Kromotografi kabineti 96
Soksilet cihazi 97
Ekstraksiyon tüpü 97
Eter ekstraksiyon cihazı 97
Babkok şişesi 98
Pergel 98
Babkok santrifüjü 98
Aliminyum kuru madde kabı 99
Desikatör 99
Kurutma dolabı 99
Vakum kurutma dolabı 100

AYIRMA ALET VE CİHAZLARI   (devam)
Nem ölçme terazisi 100
Dondurarak kurutucu 100
Keldal balonu 101
Keldal yakma ünitesi 101
Keldal distilasyon ünitesi 101
Yazıcı 102
Tartışma 103
SÜZME İŞLEMİ **104**
Huni 104
Süzgeç kağıdı 104
Cam yünü 104
Süzme erleni 105
Bühner süzgeci 105
Bir çeşit süzgeç 105
Süzgeç desteği 106
İmpinger 106
Tartışma 107
DESTİLASYON **108**
İmbik (damıtıcı) 108
Distilasyon balonu 108
Saf su cihazı 108
Mineral uzaklastırıcı (demineralizatör) 109
Flash evaporatör 109
Tartışma 110
GAZ ÖLÇÜMLERİ **111**
Respirometre 111
Tartışma 112
DENEY HAYVANLARI İÇİN ÖZEL ALETLER **113**
Kafes 113
İçecek tüpü 113
Yiyecek kabı 113
Ameliyat takımı 114
Ameliyat masası 114
Tartışma 115
MİKROSKOBİK ANALİZLER **116**
Mikroskop. 116
Lam 116
Büyüteç 116
Lens mendili 117
Mikrometre 117
Sayım haznesi 117
Boyama malzemesi 118
Doku işleme ünitesi 118
Mikrotom (kesit alma cihazı) 118
Tartışma 119
MİKROBİYOLOJİ **120**
Besiyeri 120
Dilüsyon şişesi. 120
Petri kutusu 120
Petri kutusu sterilizasyon kabı 121

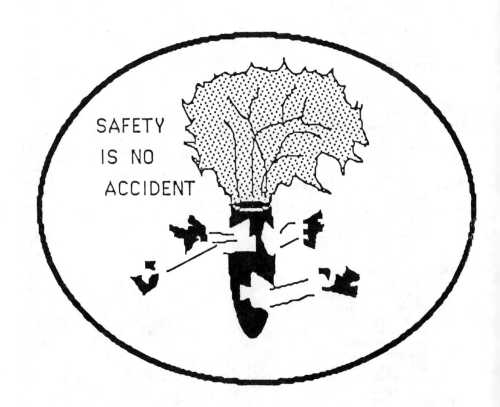

FIRST AID KIT

The **"first aid" kit** contains bandages and other emergency medical equipment.

Get me a band-aid from the **"first aid" kit.**

The adhesive tape is in the **"first aid" kit.**

FIRE EXTINGUISHER

A **fire extinguisher** is used to put out fires.

Where is the **fire extinguisher?**

Everyone should know how to use the **fire extinguisher**.

FIRE BLANKET

A **fire blanket** is a flame-resistant blanket.

A **fire blanket** is used to extinguish a person's clothing on fire.

A **fire blanket** can also be used to smother a fire on a bench or other flat surface.

SHOWER

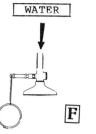

A **shower** can be used to remove chemicals from the body.

A **shower** is also effective when clothes catch fire.

Pull the chain to start the **shower.**

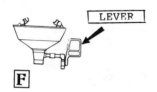

EYE WASH

An **eye wash** is used to dilute and wash away chemicals which have gotten into an eye.

Know where the **eye wash** is located so that you can find it with your eyes closed.

Be sure you know how to operate the **eye wash**.

TELEPHONE (PHONE)

Dial **phone** numbers carefully to prevent reaching a wrong number.

Write the **telephone** number of the fire department here
-------------------------------- .

Write the **phone** number of the emergency squad here
-------------------------------- .

What is the **phone** number of an ambulance?
-------------------------------- .
  a doctor?
-------------------------------- .
the police?
-------------------------------- .

WARNING LABELS, WARNING SIGNS

Warning labels or signs can alert the laboratory worker to hazards.

The **"caution"** sign means BE CAREFUL!

The **"explosive"** sign means the chemicals can be very reactive and may behave in a manner similar to a bomb.

"Biohazard" and the accompanying symbol signify a possible hazard from biological material.

RADIATION HAZARD

"Radiation hazard" and the accompanying symbol indicate the presence of radioactive material.

WEAR YOUR GOGGLES

"Wear your goggles" to protect your eyes.

DANGER

"Danger" signifies a hazardous area.

FLAMMABLE

"Flammable" indicates the possibility of fire.

CORROSIVE

Acid or alkali can be very **"Corrosive"**.

POISON

"Poison" can be fatal to people.

USE IN HOOD

"Use in hood" (use in fume cupboard) suggests appropriate ventilation is necessary.

UNDER HIGH PRESSURE

"Under high pressure" means the liquid or gas is under pressure and should be handled carefully.

CAUSTIC

"Caustic" means alkali that can burn the skin.

AVOID CONTACT WITH SKIN

"Avoid contact with the skin" means do not spill on the body.

HOT

"Hot" signifies high temperature and the ability to burn.

STERILE

If the sample contains no microorganisms, it is said to be **"Sterile"**.

KEEP IN REFRIGERATOR

"Keep in refrigerator" suggests the sample should be stored at a low temperature (less than 4 degrees centigrade) but above 0 degrees centigrade.

OUT OF ORDER

"Out of order" sign tells you the equipment is broken.

SAFETY GOGGLES

Safety goggles are worn to protect the eyes.

Wear **safety goggles** when working with hazardous material.

Where are the **safety goggles** kept?

RESPIRATOR

STRAP

Respirators are worn over the nose to protect the lungs from dust and mist.

Respirators can filter out particles but will not protect against toxic gas or vapors.

Check the filter on the **respirator.**

WHERE IS EVERYONE GOING ?

The first thing to do when working in a laboratory is to learn the location of the **safety equipment.** Find the **"First Aid Kit"** and determine what medical equipment it contains. For fire protection, you should know how to operate the **fire extinguisher** and the **fire blanket.** A **shower** can be used to extinguish clothing on fire or to dilute chemicals spilled on the body. An **"eye wash"** can be used when the chemical spill is on the face or in the eyes. A very important safety device is the **telephone** which can be used to summon help, so be sure you know how to use it and what the emergency numbers are.

Warning labels and signs can also be helpful in avoiding problems. Such labels as -- **"Caution"**, **"Explosive"**, **"Biohazard"**, **"Radiation"**, **"Danger"**, **"Flammable"**, **"Corrosive"**, **"Poison"**, and **"Caustic"** -- suggest the material should be handled carefully or the area treated with respect. **Wear goggles!** The sign **"Avoid contact with the skin"** advises you to protect your body. **"Under high and low pressure"** warns you that the liquid or gas is under abnormal or unusual pressure. **"Use in the hood"** (fume cupboard) and **"Keep in refrigeration"** suggest special handling techniques. **"Hot"** is extreme heat and **"Sterile"** is the absence of microorganisms. **"Out of order"** suggests the equipment will not work. **Safety goggles** should be placed over the eyes when washing anything with strong acids or bases and any time when there is a possibility of an explosion. **Respirators** should be used when working in dusty or misty environments.

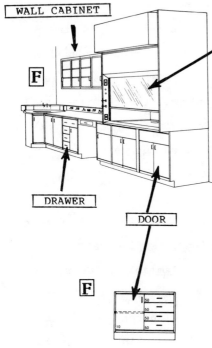

WALL CABINET

F

DRAWER

DOOR

FUME HOOD (Fume cupboard)

All chemical reactions producing vapors should be placed in a **fume hood (fume cupboard)**.

A **hood (fume cupboard)** protects you from fumes.

Volatile chemicals are handled in a **hood (fume cupboard)**.

BASE UNIT

This **base unit** has 4 drawers and 1 door.

A shelf is located in the **base unit** behind the door.

A counter (bench) top is normally placed on the **base unit** and this is usually called a laboratory bench.

F

SHELF

F

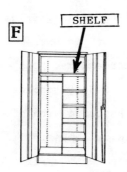

STORAGE CABINET

The **storage cabinet** has 2 doors.

Get the plastic tubing from the **storage cabinet**.

Is it in the **storage cabinet?**

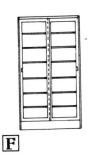

FLOOR CASE

Put the chemicals in the **floor case.**

Hand me the bottle that is in the **floor case.**

This **floor case** has 2 sliding doors.

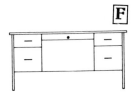

DESK

We will do our paperwork at the **desk.**

This **desk** has 5 drawers.

Place the book on the **desk.**

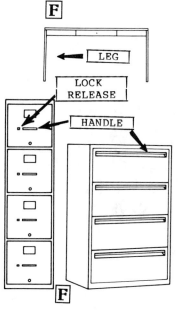

LEG

LOCK RELEASE

HANDLE

TABLE

Set the glassware on the **table.**

We will hold a conference at the **table.**

Let's sit at the **table.**

FILING CABINET

Place the laboratory data in the **filing cabinet.**

Records are kept in the **filing cabinet.**

File folders in the **file cabinet** contain the letters.

The ether extraction experiment is conducted in the
fume hood (fume cupboard) since ether is very volatile.
After the experiment is completed, the clean glassware may be
stored in the **wall cabinet, base unit, storage cabinet** or
floor case. Be sure the doors and drawers are closed
after placing the glassware in storage. The next experiment
may be set up on top of the **base unit** or on the **table.**
After this experiment is completed, be sure to record your
data at the **desk** and store the results in the **filing cabinet.**

BARBECUE IN THE HOOD ?

LABORATORY CART (Trolley)

A **laboratory (trolley) cart** is used to transport supplies.

May I borrow the **cart (trolley)**?

Where is the **laboratory cart (trolley)**?

LABORATORY CHAIR (or STOOL)

Please sit in the **chair**.

May I use the **stool** while weighing my samples?

It is not safe to stand on a **laboratory chair** or **stool**.

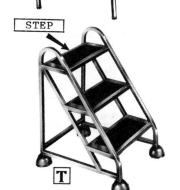

LADDER (or Steps)

Use a **ladder (the steps)** to reach the top shelf.

Be careful when using the **ladder (steps)**.

Where do we store the **ladder (steps)** when it is not in use?

Place glassware on the **laboratory cart (trolley)** so it can be taken to the laboratory bench.

Do <u>not</u> stand on the **laboratory stool** -- use the **ladder (steps)** instead!

DO YOU THINK I SHOULD USE A LADDER ?

FAUCET (Tap)

Water is transported to the lab in pipes or "mains" and arrives at the **faucet (tap)**.

The **faucet (tap)** is a valve to control the flow of water.

Turn off the **faucet (tap)**.

MIXING FAUCET (Mixer Tap)

Both hot and cold water are controlled by the **mixing faucet (mixer tap)**.

This **mixing faucet (mixer tap)** has a swivel gooseneck (spout).

Hot water is controlled by the left valve of the **mixing faucet (mixer tap)** and may be hot enough to burn your hand.

SINK

Wash the glassware in the **sink**.

Water drains (runs) from the **sink** and travels through the trap before going into the drain and then into the main sewer.

The **faucet (tap)** delivers water into the **sink**.

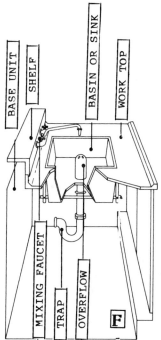

STOPCOCK

Stopcocks are used to control the flow of water, gas, or air.

Stopcocks often control the main supplies of water, gas or air and may be used for servicing or emergency use.

A label on a **stopcock** often tells you what the valve controls.

COLD WATER

A **cold water** stopcock controls the flow of cold (unheated) water.

HOT WATER

A **hot water** stopcock controls the flow of **hot water** and the water may be hot enough to burn your hand.

DISTILLED WATER

Use **distilled water** in this reaction because it has been distilled and contains <u>no</u> minerals.

STEAM

Steam is water that has been changed to a hot gaseous state (will burn skin rapidly).

AIR

Air is often used to drive stirring devices or to cool hot materials.

GAS

The burner is connected to the **gas** stopcock.

OXYGEN (O$_2$)

Oxygen is mixed with gas to produce a hotter flame.

NITROGEN (N$_2$)

Nitrogen is often used as a packaging environment to exclude oxygen and prevent oxidation.

VACUUM [Less than normal (atmospheric) pressure]

Vacuum is sometimes used to aid or accelerate filtering.

NEEDLE VALVE

A **needle valve** is used when sensitive control of flow rate is desired.

Turn off the **needle valve.**

Connect a **needle valve** to the gas line.

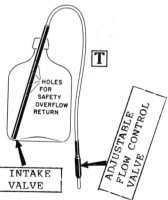

HOLES
FOR
SAFETY
OVERFLOW
RETURN

INTAKE
VALVE

ADJUSTABLE FLOW CONTROL VALVE

SIPHON

A **siphon** is used to transfer liquid from a higher level to a lower level.

Siphon the distilled water into the beaker.

How do you start the **siphon?**

A **faucet (tap)** may deliver **hot** or **cold water** into a **sink**. A **mixing faucet (mixer tap)** can mix **hot** and **cold water** to produce warm water for washing glassware.

Stopcocks can be used to control a variety of liquids or gases. In addition to **hot** and **cold water,** the **stopcock** might regulate **distilled water** or even **steam**. Gases such as **air,** burner **gas** (flammable), **oxygen** or **nitrogen** are also often controlled by **stopcocks**. A lack of air or **vacuum** can also be controlled by these valves. If finer control is desired, a **needle valve** is often used. If liquid has to be moved from a higher to a lower level and it is not under pressure, a **siphon** is often used.

I CAN FIX IT WITH A HAMMER !

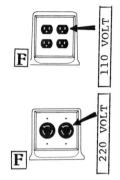

ELECTRICAL OUTLET (Electrical Socket)

There are four 110-volt **electrical outlets (sockets)** in this box.

Plug the motor into the **electrical outlet (socket)**.

There are two 220-volt **electrical outlets (sockets)** in this box.

The plug has to be rotated prior to removal from this **electrical outlet (socket)**.

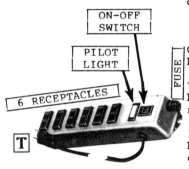

OUTLET BOX OR STRIP [Distribution board (box)]

This **outlet strip (distribution board)** may be moved to the area needed.

In this **outlet strip (distribution box)**, there are six 110-volt **outlets,** a pilot light, an on-off switch and a fuse.

The six 110-volt **outlets** in this **strip (board)** do not have a ground (earth) wire connection; therefore, the equipment will not be grounded (this may be dangerous).

TRANSFORMER OR POWERSTAT ("Variac")

A variable **transformer (Variac)** can be used to control voltage from 0 to 140 volts.

Adjust the **transformer (Powerstat)** to deliver 60 volts.

Plug the heating element into the **transformer (Variac)**.

TERMINALS

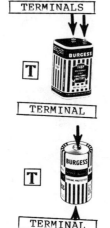

TERMINAL

TERMINAL

BATTERY

A **battery** is used to store electrical energy.

Batteries supply direct current (d.c.).

Batteries are labeled with the amount of voltage they supply.

This shape of **battery** is often referred to as a flashlight **battery** or a "dry" **battery**.

AMPERE METER

110 VOLT PLUG

BATTERY CONNECTIONS

BATTERY CHARGER - for acid batteries and only certain rechargeable "dry" batteries (e.g., nickel-cadmium cells)

A **battery charger** is used to recharge a battery.

A **battery charger** converts alternating current (a.c.) into direct current (d.c.).

Some batteries cannot be charged and must <u>not</u> be connected to a **battery charger**.

The normal American voltage is 110 volts a.c. and can be obtained from **outlets (sockets)** that have 2 parallel openings. If there is a D-shaped opening below and between these parallel openings, this is a ground (earth) connection. A 220-volt **outlet (socket)** has 3 curved openings arranged in a circle. The **electrical outlets (sockets)** may be located in the wall, in a box, on the shelf support above the lab bench or in a portable **outlet strip (distribution box)**. **Variable transformers (Variac)** are used when adjustable voltages are needed. **Batteries** are used to store and supply direct current (d.c.) voltage and the **battery** is selected depending upon the voltage desired. After the **battery** has been used, it will become discharged but some can be recharged by the use of a **battery charger.**

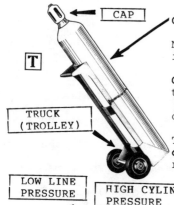

CAP

TRUCK
(TROLLEY)

GAS CYLINDER or TANK

Many types of gases may be obtained in **gas cylinders (tanks)**.

Gas cylinders (tanks) are transported by a gas cylinder truck (trolley) equipped with a safety chain.

The cap is removed from the gas **cylinder (tank)** and a pressure regulator is attached.

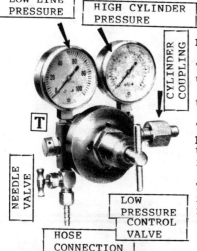

LOW LINE PRESSURE

HIGH CYLINDER PRESSURE

CYLINDER COUPLING

NEEDLE VALVE

LOW PRESSURE CONTROL VALVE

HOSE CONNECTION

PRESSURE REGULATOR

The **pressure regulator** is used to reduce the cylinder (tank) pressure and control the gas flow.

The high cylinder (tank) **pressure gauge** indicates the amount (pressure) of gas in the cylinder (tank).

The low line **pressure gauge** indicates pressure of gas flow.

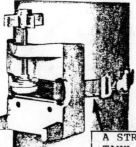

GAS CYLINDER CLAMP

A **gas cylinder clamp** is used to secure the cylinder (tank) to a solid support such as a bench or wall.

For safety purposes all gas cylinders (tanks) should be secured by **gas cylinder clamps**.

A STRAP HOLDS THE TANK OR CYLINDER TO THE CYLINDER CLAMP

Gases such as oxygen, nitrogen, carbon dioxide, air, nitric oxide, helium, hydrogen and many others are often available in **gas cylinders (tanks)**. These gases range from very dangerous, if inhaled, through flammable to inert. If the valve on the end of the cylinder (tank) is accidently broken off, all gases are dangerous since the rapid release from the cylinder (tank) will propel the cylinder (tank) violently in the opposite direction. For this reason, a cap is kept on the **cylinder (tank)** when it is in transit or not in use. **Cylinders (tanks)** are often transported in the laboratory by a 2-wheeled truck (trolley) and secured by a safety chain. When placed in a laboratory, the **cylinder (tank)** is secured to a stationary object with a **gas cylinder clamp** to prevent falling. To control the release of gas from the **cylinder (tank)**, a **pressure regulator** is attached to the **gas cylinder (tank)**. **Pressure regulators** are usually not interchangeable for different types of gases. To prevent their use with the wrong types of **cylinders (tanks)** there are several types of screw threads used on both **cylinders (tanks)** and **regulators**.

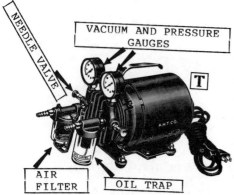

VACUUM AND PRESSURE PUMP

This **pump** can supply a source of air under pressure.

This **pump** can also create a vacuum.

This **vacuum and pressure pump** is portable.

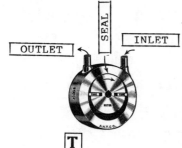

VANE PUMP MOVEMENT

The rotor contains 2 spring-loaded **vanes.**

The **vane pump** contains a crescent-shaped air chamber.

This **vane pump** rotates in a clockwise direction.

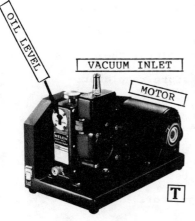

VACUUM PUMP (Rotary Oil Pump)

This **vacuum pump** can be used for vacuum distillation.

The motor is connected to the **vacuum pump** by a V-belt.

This **vacuum pump** contains oil and its level should be maintained.

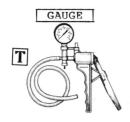

HAND PUMP

This **hand pump** can produce a pressure or a vacuum.

This **hand pump** operates with a finger squeeze-type handle.

This light-weight portable **hand pump** can be used when small volumes are required.

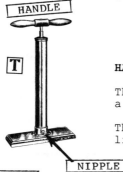

HAND PUMP (floor model)

This **hand pump** can also produce a pressure or a vacuum.

This **hand pump** looks very much like an automobile tire pump.

ASPIRATOR (Water Pump)

The **aspirator (water pump)** produces a vacuum with the aid of flowing water.

Aspirators (water pumps) are made of metal or of plastic.

Six to nine liters (litres) of water per minute are required (pass through) by an **aspirator (water pump)**.

Vacuum and pressure or air pumps are available in
a variety of styles. The motor driven pumps are available in
a vane type or in a piston type. The hand pumps
of some models can be driven by squeezing a trigger-type
device and, in others--by pushing and pulling a plunger
attached to a handle. With the aspirator-type (water
pump), water is forced through the device and a vacuum is
produced.

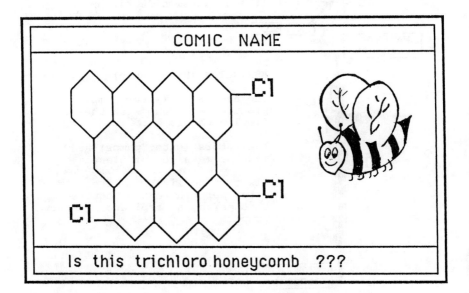

COMIC NAME

Is this trichloro honeycomb ???

BELOW ZERO DEGREES CENTIGRADE

ABOVE ZERO DEGREES CENTIGRADE

DOOR

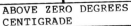

T   REFRIGERATOR-FREEZER

Most **refrigerators** maintain cold temperatures above freezing and have a **freezer** section that maintains temperatures below freezing.

Freezers usually have only one compartment that maintains temperatures below 0 degrees centigrade.

Volatile chemicals placed in a refrigerator may often cause an explosion, unless the **refrigerator** is fitted with sealed electrical switches.

DRY ICE (Solid Carbon Dioxide) STORAGE CONTAINER

T   TOP

Dry ice (solid carbon dioxide: Care! Temperature -78 degrees centigrade) **storage containers** are insulated chests used for storage of **dry ice.**

Dry ice (solid carbon dioxide) can freeze the skin and must be handled with gloves.

Insulated **storage containers** retard the loss (sublimation) of **dry ice (solid carbon dioxide).**

Do not use **dry ice (solid carbon dioxide)** in unventilated spaces.

DRY ICE BOX (Solid Carbon Dioxide Storage or Shipping Container)

Dry ice (solid carbon dioxide) boxes are often made of polystyrene.

Dry ice (solid carbon dioxide) boxes can be used for shipment or storage of perishables in a frozen state.

Insulated **boxes** can also be used to keep samples hot.

DEWAR FLASK (Thermos Flask)

A **Dewar flask (thermos flask)** is a double-walled glass container with a vacuum between the walls.

A **Dewar thermos flask (thermos flask)** can be used to maintain a hot or cold temperature.

The silvered and evacuated glass **Dewar flask (thermos flask)** is well insulated.

Cold storage may be maintained in the laboratory in a variety of ways. A **refrigerator** maintains temperature above freezing. Below freezing temperatures may be found in the **freezer section** of a **refrigerator** or in a **freezer**. A **freezer** may be an upright model with a vertical door or a chest-type with the door on top. Insulated **containers** are also used to maintain temperature. They may be found in heavy laboratory **dry ice (solid carbon dioxide) boxes** used for storage or shipping. **Dewar flasks (thermos flasks)** are often used to maintain a cold temperature around a cold trap in a distillation unit.

A HOUSE FOR YOUR SNOWMAN !

SPATULA

A **spatula** is used to pick up a chemical sample.

This **spatula** has a steel blade.

Where do you keep the **spatula**?

HANDLE

SCOOP

A **scoop** is used to transfer larger samples.

This **scoop** is made of aluminum (aluminium).

A **scoop** is a big spoon.

TONGS

Tongs are used to move small items of laboratory equipment.

Use **tongs** to handle hot equipment or to handle items containing hazardous chemicals.

There are many styles of **tongs** to handle equipment of different shapes.

POLICEMAN

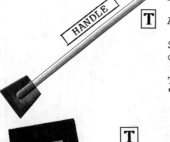

HANDLE

A **"policeman"** is used for stirring.

Scrape the sample from the container with the **"policeman"**.

The thin rubber blade of the **"policeman"** is flexible.

NOTEBOOK

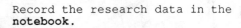

Record the research data in the **notebook**.

The **notebook** contains all the results of the experiment.

Label all items in the **notebook**.

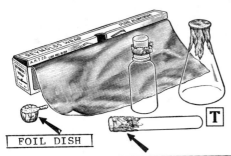

| FOIL DISH |

| TEST TUBE AND FOIL CAP |

FOIL

Aluminum **foil** is used in capping glassware.

You can shape a disposable weighing dish from **foil**.

Foil is packed in rolls.

BAG

Bags are used to store samples.

Polyethylene (polythene) **bags** are transparent.

Disposable **bags** are convenient sample containers.

EVAPORATING DISH

Porcelain **evaporation dishes** are placed in the oven.

Place the sample to be evaporated in this **dish**.

Evaporation dishes are available in many sizes.

APRON

Use the **apron** to protect your clothes.

Some **aprons** are disposable.

Clean **aprons** are located in the cabinet.

ASBESTOS GLOVES

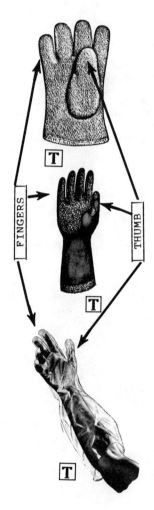

Use **asbestos gloves** when handling hot objects.

The crucible will burn your fingers unless you use **asbestos gloves**.

Asbestos gloves are good heat insulators.

RUBBER GLOVES

Use **rubber gloves** when washing glassware.

Rubber gloves are useful when transferring corrosive chemicals.

Rubber gloves protect your hands.

Do <u>not</u> use **rubber gloves** with hot objects.

DISPOSABLE GLOVES

Disposable gloves protect against solvents, acids and detergents.

For bacterial protection, use **disposable gloves**.

Disposable gloves allow a great deal of manual dexterity.

Do <u>not</u> use **disposable gloves** with hot objects.

WASH BOTTLE

By squeezing the **wash bottle**, a stream of liquid is produced.

Wash bottles are available in many sizes.

Water as well as solutions or solvents can be delivered from a **wash bottle**.

DROPPING BOTTLE

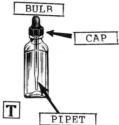

Dropping bottles are often used for reagents.

One drop can be delivered from a **dropping bottle.**

The rubber bulb controls the flow from the **dropping bottle.**

RUBBER BULB

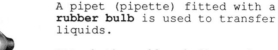

A pipet (pipette) fitted with a **rubber bulb** is used to transfer liquids.

Attach the **rubber bulb** to the pipet (pipette).

Always use a **rubber bulb** on a pipet (pipette) to avoid liquids coming in contact with your mouth.

BRUSH

Use the **brush** for cleaning.

Brushes are available in many shapes.

Wash the glassware with the **brush.**

CLEANING COMPOUND

A detergent is a **cleaning compound.**

Cleaning compounds may be solid, powder or liquid.

Cleaning compounds are an aid in washing dirt from equipment.

TISSUE

Kleenex **tissue** is absorbent.

Special **tissue** is used on optical lens.

Use a paper towel rather than a **tissue** on large spills.

WALL CLOCK

A **wall clock** is used to tell time.

This **wall clock** indicates it is 12 minutes till (to) 5 (or, four forty-eight).

Wall clocks in the U.S. have a 12-hour cycle and p.m. indicates time in the afternoon (4:48 p.m. = 16:48 on a 24-hour clock).

INTERVAL TIMER

An **interval timer** is used to measure the time of a chemical reaction.

Pull the lever down to start this type of **interval timer**.

An alarm in the **interval timer** will sound when the set time has expired.

TIME SWITCH

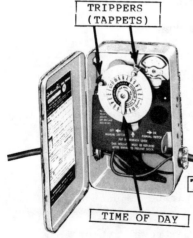

A **time switch** is used to start and stop electrical equipment automatically.

The **time switch** will repeat "on" and "off" cycles every 24 hours.

Trippers (Tappets) in the **time switch** can be adjusted for different time cycles.

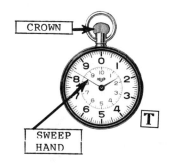

STOPWATCH

Use a **stopwatch** to accurately time an event.

The crown is used to wind the **stopwatch.**

The **stopwatch** crown is also used to stop, start and reset the timer.

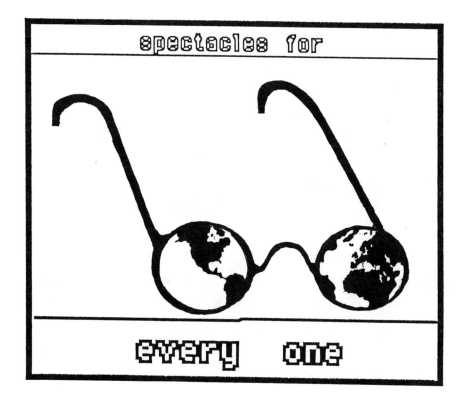

A great deal of **hand laboratory equipment** is located in almost every laboratory and a few examples are as follows. A **spatula** is used to transfer small quantities of a sample but a **scoop** is necessary when dealing with larger quantities. Tweezers or forceps are used to handle very small items but **tongs** are necessary for handling larger pieces of equipment. A **"policeman"** can be used as a stirring rod but it is also helpful when scraping a sample from a container.

Laboratory data should always be recorded in a laboratory **notebook**. **Foil** is useful to make caps or containers or to wrap samples. **Bags** are also used as sample containers. An **evaporating dish** can be heated to drive moisture from the sample. Clothing and the body should be protected with an **apron** and the hands should be protected with **gloves**. **Asbestos, rubber, and disposable gloves** all offer different types of protection. Liquid may be dispensed by a **wash bottle** or a **dropping bottle** or with the aid of a pipet (pipette) and a **rubber bulb**. Dishes may be washed with a **cleaning compound** and a **brush**. A small liquid spill may be absorbed with a **tissue** but disposable paper towels are necessary for larger volumes of liquid.

Timing devices take many forms and the **wall clock** is used to tell the time of day. An **interval timer** tells you when to stop a reaction; a **time switch** starts or stops equipment automatically; and, a **stopwatch** is used when accuracy is required.

PLIERS

Pliers are used to hold or turn nuts.

These **pliers** have a pipe wrench area.

Wire can be cut with one section of these **pliers**.

NEEDLE NOSE PLIERS (Fine Nose Pliers)

Needle (fine) nose pliers are used to handle small items.

These **needle (fine) nose pliers** also have a wire cutting section.

Needle (fine) nose pliers are used to bend wire.

WIRE CUTTER PLIERS

Wire cutter pliers are useful in cutting wire in tight places.

These **wire cutting pliers** are of the diagonal cutting type.

Do not cut heavy wire with these **wire cutter pliers**.

TIN SNIPS

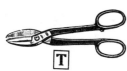

Tin snips are used to cut metal.

Wire can be cut with **tin snips**.

Use the **tin snips** to cut the wire screen.

SCISSORS, OR SHEARS (Normally heavier)

Shears or **scissors** are used to cut paper.

Cardboard can also be cut with **shears** or **scissors**.

Cut the paper with the **scissors** or **shears**.

CRESCENT WRENCH (Adjustable Wrench)

A **crescent wrench** will adjust to fit a square or a hex head nut.

A **crescent wrench** is often called an adjustable wrench.

Use the **adjustable wrench** to tighten the nut.

OPEN-END WRENCH

Select the **open-end wrench** to fit the nut.

The **open-end wrench** is a different size on each end.

A **closed-end wrench** fits around the nut but the **open-end wrench** fits only 2 sides of the nut.

HEX WRENCH SET

Use a **hex wrench** to turn the setscrew.

A **hex wrench** may be part of a set or may be an individual wrench.

Hex wrenches are available in many sizes.

HAMMER

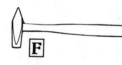

A **hammer** is used for driving nails.

Don't hit your thumb with the **hammer**.

Break the sample with a **hammer**.

SCREWDRIVER

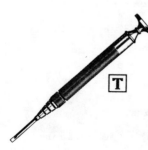

A **screwdriver** is used to turn screws.

A regular **screwdriver** fits a straight slot.

A **screwdriver** that fits a cross-type slot is called a Phillips **screwdriver**.

DRILL BIT

A **drill bit** is used to bore a hole.

Different sized **drill bits** are available.

Place the **drill bit** in the drill.

METAL FILE

Metal files are available in many shapes.

A **metal file** is used to shape metal.

A **metal file** is also used to etch glass.

VISE (Vice)

A **vise (vice)** is used to clamp items.

This **vise (vice)** also contains a small anvil.

Clamp the **vise (vice)** to the bench.

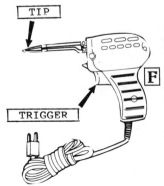

TIP

TRIGGER

SOLDERING GUN

A **soldering gun** is used to melt solder.

A **soldering gun** that is a large pencil shape is called a soldering iron.

Be careful of the hot tip on the **soldering gun.**

Many tools are used in the laboratory. **Pliers** are
frequently used.   They may be regular **pliers** or **wire
cutter pliers.   Tin snips, shears** or **scissors** are
used for cutting.   Items are turned with **wrenches** such as
an **adjustable** or **crescent wrench,** an **open-end
wrench,** a **closed-end wrench** or a **hex wrench.** A
hammer is used for driving or pounding. A **screwdriver**
is a very useful laboratory tool and the two most popular
types are regular and Phillips. Holes are drilled with
drill bits and metal is shaped with a **metal file.** A
vise (vice) is used to hold items and a **soldering gun**
is used to solder wire joints.

IS THE GLASS STOPPER STUCK ?

TAPE DISPENSER

Transparent tape is placed in the **tape dispenser**.

The **tape dispenser** has a cutter for cutting the tape.

Masking tape can be obtained from the **tape dispenser**.

LABEL MAKER

A **label maker** embosses plastic tape.

Type the label on a **label maker**.

Squeeze the handle of the **label maker** to emboss a letter on the tape.

LABEL

Write on the **label**.

The **label** is used to identify the sample.

Some **labels** are gummed.

BALL-POINT PEN

Use the **ball-point pen** for marking.

Ball-point pens are available in many ink colors.

Mark the label with the **ball-point pen**.

WAX PENCIL

Use the **wax pencil** to write on glass.

Some **wax pencils** are heat resistant.

Rolled paper surrounds the wax in a **wax pencil**.

FELT TIP MARKING PEN

Keep the cap on the **felt tip marking pen** when it is not in use.

Felt tip marking pens are available in many colors.

Most **felt tip marking pens** are non-refillable.

RULE OR RULER

Measure the sample with the **rule.**

The **ruler** is graduated in inches and centimeters.

Length is measured with a **ruler.**

CALCULATOR

Add the numbers on the **calculator.**

Calculate the percentage on the **calculator.**

Check the math (maths) on the **calculator.**

Office hand tools are also useful in the laboratory.
A tape dispenser can be used to dispense a variety of
types and widths of tape which is normally used for sealing
packages. A label maker is a very useful item for
printing labels so that everyone will know where items are
stored. Gummed labels are also useful for identification.
Writing instruments are available in many types and would
include a pencil, ball-point pen, wax pencil and felt
tip marking pen. A rule or ruler is used to
measure linear distances. The calculator has almost
replaced the slide rule as a method of figuring mathematics.

TAPE DISPENSER WENT WILD !

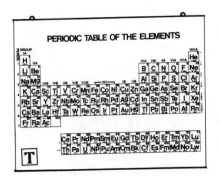

PERIODIC TABLE OR CHART

A **periodic table or chart** lists all of the elements.

The atomic symbol, atomic number and atomic weight may be found in the **periodic table.**

The **periodic table** groups the elements into families.

ARE YOU SURE THIS IS WHERE YOU CAN GET A PERIODIC TABLE ?

The **periodic table or chart** is very useful in the laboratory. It places atoms which behave similarly into groups. The **periodic chart** also supplies a great deal of useful information about the elements.

HE LOOKED UP THE PROPERTIES OF HE IN THE PERIODIC CHART.

ACETIC ACID CH_3COOH

    Glacial **acetic acid** is 99.5% **acetic acid.**

ALCOHOL CH_3CH_2OH

    The term **alcohol** refers to ethyl **alcohol.**

AMMONIUM HYDROXIDE NH_4OH

    Ammonium hydroxide is 29% NH_3 in water solution.

CALCIUM HYDROXIDE $Ca(OH)_2$

    Lime water is a solution of **calcium hydroxide.**

HYDROCHLORIC ACID HCl

    Concentrated **hydrochloric acid** contains 38% **HCl.**

NITRIC ACID HNO_3

    Concentrated **nitric acid** contains 70% **HNO_3.**

POTASSIUM HYDROXIDE KOH

    Potassium hydroxide is sometimes called caustic potash.

SODIUM HYDROXIDE $NaOH$

    Sodium hydroxide is sometimes called caustic soda.

SULFURIC ACID H_2SO_4
(SULPHURIC ACID)
    Concentrated **sulfuric (sulphuric) acid** contains 93-98%
    H_2SO_4.

REAGENT LABEL

METRICS. Distinctive solid
black bar readily identifies
metric packaging.

PURITY GRADE
Certified = highest
Fisher Grade.
ACS = meets reagent
specifications of
American Chemical
Society, the
universal authority.

SIZE stated
in both systems
of measurement.

**CATALOG
NUMBER**

LOT
ANALYSIS

Certificate of Actual Lot Analysis
$CH_3CO.CH_3$ F.W. 58.08
Color (APHA)
Density (g/ml) at 25°C 0.7857
Boiling Point .... 56.1°C ± 0.1°C
Boiling Range
(First drop to drypoint) . 55.9°–56.3°C
Assay (CH_3COCH_3) 99.5%
Residue after evaporation 0.0002%
Solubility in water .... P.T.
Acidity (as CH_3COOH) 0.002%
Alkalinity (as NH_3) 0.0002%
Aldehyde (HCHO) 0.0005%
Methanol (CH_3OH) 0.05%
Substances reducing Permanganate . P.T.
Water (H_2O) 0.5%
Isopropyl Alcohol 0.01%

**EXTREMELY
FLAMMABLE
DANGER!**

FLASH POINT .... 0°F

Keep away from heat, sparks and open
flames.
Keep container closed.
Use with adequate ventilation.
Avoid prolonged or repeated contact with
skin.
KEEP OUT OF REACH OF CHILDREN

A-18

4 L
(1.06 GAL.)

LOT
783447

Class 1B

Certified A.C.S.

Acetone

For laboratory and
manufacturing use
only, not for drug, food,
or household use.

FISHER SCIENTIFIC COMPANY
Chemical Manufacturing Division
Fair Lawn, New Jersey 07410
Made in U.S.A.

SAFETY
DATA

LOT
NUMBER

WHERE APPLICABLE; OSHA
CLASSIFICATION FOR
FLAMMABLE AND COMBUSTIBLE
LIQUIDS.

THE SOURCE

Many classes of **reagents** are found in the laboratory.
For example, **acids** such as **acetic, hydrochloric, nitric**
and **sulfuric (sulphuric)** are very common. **Bases** such as
ammonium, calcium, potassium and **sodium hydroxides** are
also available. Many types of **alcohol** are used in the
laboratory but if only the word **"alcohol"** is used, it refers
to **ethyl alcohol**. **Reagents** arrive at the lab in bottles or
cans and they are always **labeled**. The **label** may indicate
such things as the **analysis** of the **reagent**, the **size**
or **quantity** of the **reagent**, the **purity** of the **reagent**,
the **classification** of the **reagent**, the **lot number** of
the **reagent** and the **safety data** pertinent to the **reagent**.

ADD ACID TO WATER !

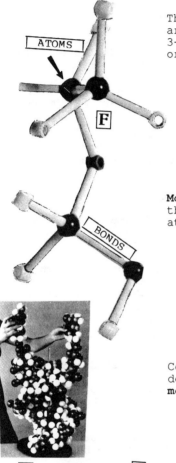

Three types of **molecular models**
are shown on this page and they are
3-dimensional representations of
organic compounds.

Molecular models not only show
the atoms but the bonds between the
atoms.

Color (colour) coded atoms of a
definite scale are used in
molecular models.

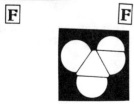

Molecular models are helpful in
understanding the 3-dimensional
configuration of organic compounds.

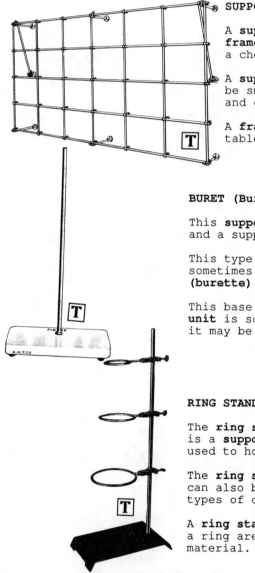

SUPPORT OR FRAME

A **support unit** or
frame can be used to mount
a chemical apparatus.

A **support** or **frame** can
be small and simple or large
and complex.

A **frame** can sit on the
table or mount on the wall.

BURET (Burette) STAND

This **support unit** has a base
and a support rod.

This type of **support unit** is
sometimes called a **buret
(burette) stand**.

This base unit of a **support
unit** is sometimes porcelain or
it may be enameled steel.

RING STAND (Retort Stand)

The **ring stand (retort stand)**
is a **support unit** that is often
used to hold rings.

The **ring stand (retort stand)**
can also be used to hold other
types of clamps.

A **ring stand (retort stand)** and
a ring are useful when heating
material.

TRIPOD

A **tripod** can serve some of the same functions as a **ring stand (retort stand)**.

A **tripod** is not adjustable in height.

This **tripod** is equipped with removable concentric rings.

BURET CLAMP (Burette Clamp)

A **buret (burette) clamp** is attached to a vertical support rod.

The **buret (burette) clamp** holds 2 burets (burettes).

The graduations on the buret (burette) are not obscured by a **buret (burette) clamp.**

THUMBSCREW (BOSS HEADS)

CLAMP HOLDER

A **clamp holder** which may hold a clamp is attached to a support.

Clamp holders are available in many styles.

A thumbscrew (boss head) is used to secure the **clamp holder** to the support.

CLAMP

A **clamp** is held by a **clamp holder**.

This **clamp** has 2 adjustable jaws.

Small and medium size objects can be held with this type of **clamp.**

CLAMP WITH HOLDER

CHECKNUT

THUMBSCREW
(BOSS HEADS)

Clamp with holder is a **clamp** and **clamp holder** in one unit.

This **clamp and holder** has a stationary jaw and a movable jaw.

The checknut allows the **clamp** to be rotated in relation to the **holder**.

A SUPPORT OR FRAME IS HELPFUL.

A **support** or **frame** is made from vertical and
horizontal bars. Small versions are vertical bars attached to
a base. These may be called **support units** or **ring
stands (retort stands)**. **Rings** are often attached to
the **ring stand (retort stand)**. A **ring** with 3 legs
attached is called a **tripod**. A **buret (burette) clamp**
attaches a buret (burette) to a **support** rod. **Clamp
holder** and **clamps** or a **clamp with holder** attaches
objects to **support units**.

SHOULD I BE USING A CLAMP HOLDER ?

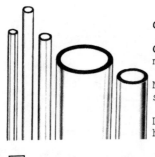

GLASS TUBING

Glass tubing is available in many sizes.

Most **glass tubing** can be softened in a gas-air flame.

Different types of **glass tubing** have different degrees of hardness.

CAPILLARY TUBING

Glass tubing with a small internal bore (hole) is called **capillary tubing.**

Capillary tubing can be used to make capillary pipets (pipettes).

Melting points are often determined in very thin **capillary tubing.**

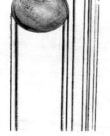

GLASS ROD

A stirring rod can be made from a **glass rod.**

Glass rods are often made from soft glass.

A **glass rod** does <u>not</u> have a hole in the center.

BALL AND SOCKET JOINT

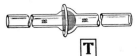

A **ball and socket joint** is used
to join glass tubing.

A **ball and socket joint** 12/2
indicates a ball of 12 mm in
diameter and a bore of 2 mm.

A **ball and socket joint** is
useful when alignment is difficult.

O-RING JOINT

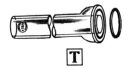

An **O-ring joint** provides a rigid,
greaseless, vacuum tight seal.

The glass tubing is joined by an
O-ring joint.

Place the **O-ring** between the
O-ring joint.

STANDARD TAPER GROUND GLASS JOINT ("Quickfit")

A **standard taper ground
glass joint (quickfit)** of
24/40 indicates a 24 mm
diameter at the widest point
and 40 mm long.

The **standard taper ground
glass joint (quickfit)** is
often used to connect a flask
to a condenser.

Springs are used on some
**standard taper ground glass
joints (quickfits)** to hold
them together.

PINCH CLAMP

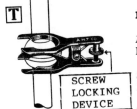

SCREW
LOCKING
DEVICE

A **pinch clamp** can be used on a
ball and socket and an O-ring joint.

Some **pinch clamps** have a screw
locking device.

The **pinch clamp** has a spring
closed, forked jaw.

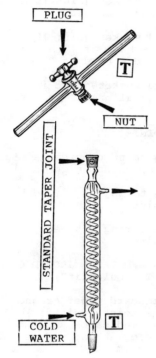

GLASS STOPCOCK

A **glass stopcock** may have a teflon or glass plug.

A **glass stopcock** is used to control flow.

A nut often holds the plug in the **stopcock.**

CONDENSER

A **condenser** cools a vapor and changes it into a liquid.

Cold water is circulated through the **condenser.**

The **condenser** contains a condensing tube surrounded by a water jacket.

TEST TUBE

A **test tube** is an elongated glass container often used for chemical reactions.

Heat the sample in the **test tube.**

There are many sizes of **test tubes.**

TEST TUBE HOLDER

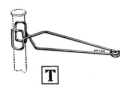

A **test tube holder** is used when heating the tube.

The **test tube holder** has self-closing jaws.

A spring keeps the jaws closed on the **test tube holder.**

TEST TUBE RACK

Test tube racks are made of wood, wire, plastic or metal.

This **test tube rack** holds 12 tubes.

This **test tube rack** does not have drying pins.

WATCH GLASS

The **watch glass** will allow steam to escape from the beaker.

Evaporate the sample in the **watch glass.**

Watch glasses are available in many sizes.

SAFETY HAND GRIP

Use the **safety hand grip** when pushing glass tubing through a stopper.

When breaking glass tubing, use the **hand safety grip.**

The **safety hand grip** can handle glass tubing of different sizes.

GLASS TUBING CUTTER

The **glass tube cutter** is used
for scoring the tube.

Some **glass tube cutters** contain
a hot wire.

After cutting the tube with the
glass tube cutter, the cut end
of the tube should be polished in a
flame.

A GLASS TUBING CUTTER WORKS BETTER.

Glass does not react with most chemicals and, therefore, is used in the chemical laboratory. **Glass tubing, capillary tubing** and **glass rods** are the most common types of glass pieces used in the lab. **Glass tubing** is often joined with **ball and socket, O-ring** or **standard taper ground glass (quickfit) joints.** The **joints** are held together by gravity, **springs** or **pinch clamps. Stopcocks** are used to control flow through **tubing** or **containers.** A **condenser** is tubing fitted with a water jacket. **Test tubes** are small containers and a **test tube holder** is used when heating the tubes. To keep the tubes in order and hold them upright, they are often placed in a **test tube rack.** A **watch glass** is often placed above a beaker to condense some of the boiling material and return it to the beaker. **Hand safety grips** should be used when holding **glass tubing** and a **glass tubing cutter** is used to etch (score) glass prior to breaking.

I HOPE YOU USED QUICKFIT JOINTS !

SPOUT
(LIP)

BEAKER

A **beaker** may or may not have a spout (lip).

Beakers are made of glass, plastic or steel.

Many sizes of **beakers** are used.

BOTTLE

Store the chemical in the **bottle**.

Bottles may be clear or amber or various other colors.

Bottles are made of glass or plastic.

ERLENMEYER FLASK (Conical Flask)

An **Erlenmeyer (conical) flask** is a convenient container for mixing chemicals.

The graduations on an **Erlenmeyer (conical) flask** are only approximate.

Some **Erlenmeyer (conical) flasks** have a standard taper joint on the top.

HANDLE
OR BAIL

PAIL OR BUCKET

Pails or **buckets** are made from metal or plastic.

A large container with a handle is called a **pail** or **bucket**.

Pour the liquid into the **bucket** (or **pail**).

FLORENCE FLASK

Flat bottom **Florence flasks** are used in some experiments.

This **Florence flask** has a standard taper joint but some do not.

Boil the sample on a hot plate in the **Florence flask**.

VIAL

A **vial** is a small container.

Store the sample in the **vial**.

This **vial** has a cap but some are closed with a stopper.

CAP

Caps are made of plastic or metal.

Place the **cap** on the bottle.

Most **caps** have a cardboard or foil liner.

STOPPER (Rubber or Plastic Stopper)

A **stopper** may be solid or contain holes.

The larger the **stopper** number (e.g., #6), the larger the **stopper**.

Close the flask with the **stopper**.

GLASS STOPPER

This **glass stopper** has a penny head.

The standard taper size of a **glass stopper** indicates the diameter of the ground zone at the widest point.

Place the **glass stopper** in the flask.

CORK

A **cork** is also used as a stopper.

A **cork** is made from the bark of a tree.

T

The larger the **cork** number, the larger the **cork**.

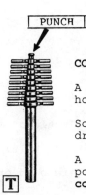

PUNCH

CORK BORER

A **cork borer** is used to cut holes in a stopper.

Some **cork borers** are motor driven.

T

A solid cork punch pushes the cut portion of the stopper from the **cork borer**.

OUR SPEAKER IS AN EXPERT ON CHEMICAL CONTAINERS !

Many types of **containers** are used in the chemical laboratory and a **beaker** is a very popular container. Reagents are usually stored in **bottles** and small samples are stored in **vials**. **Erlenmeyer (conical) flasks** and **Florence flasks** are often used to contain chemical mixtures which are to be heated. A **pail** or **bucket** is used to contain large quantities of materials. Containers are closed with **caps, stoppers, glass stoppers** or **corks**. A **cork borer** is used to cut holes in corks or rubber stoppers.

THIS IS A LABORATORY BEAKER ?

RUBBER TUBING

Rubber tubing is used to transfer liquids or gases.

Rubber tubing is available in many bore sizes and wall thicknesses.

Strong, elastic **rubber tubing** provides a gas-tight connection.

PLASTIC TUBING

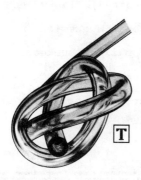

Plastic tubing made from tygon or teflon is very chemical resistant.

Polyethylene (Polythene) **plastic tubing** like **rubber tubing** is available in many bore and wall thickness sizes.

Plastic tubing is usually transparent.

TUBING CLAMP

A **tubing clamp** is used to stop the flow in tubing.

Regulate the flow with a **tubing clamp**.

Stop the flow with a **tubing** clamp.

Rubber or **plastic tubing** is used to transfer
liquids and gases in the laboratory. Flow in the tubing is
controlled by a **tubing clamp.**

A TUBING CLAMP TO CATCH AN OCTOPUS?

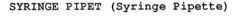

SYRINGE PIPET (Syringe Pipette)

Very small volumes are measured in a **syringe pipet (pipette)**.

A **syringe pipet (pipette)** is often used to inject a sample into a chromatograph.

Some **pipets (pipettes)** have fixed and some have removable needles.

Precision of a micro **syringe pipet (pipette)** is within ± 1%.

PLUNGER

T

TRANSFER OR VOLUMETRIC PIPET (PIPETTE)

A **transfer** or **volumetric pipet (pipette)** is used to accurately measure a specific volume.

This is a 10 ml **transfer** or **volumetric pipet (pipette)**.

Measure 50 ml with the **transfer** or **volumetric pipet (pipette)**.

T

MEASURING OR GRADUATED PIPET (Measuring or Graduated Pipette)

Measure 3.4 ml with the **measuring** or **graduated pipet (pipette)**.

The last 2 ml of this 25 ml **measuring pipet (pipette)** is not graduated.

Each graduation is 2/10 ml on this **measuring pipet (pipette)**.

T

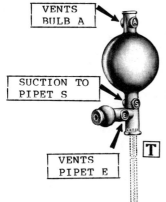

VENTS BULB A

SUCTION TO PIPET S

VENTS PIPET E

PIPET FILLING ATTACHMENT OR BULB (Pipette Filling Attachment or Bulb)

Use a **pipet (pipette) filling bulb** for toxic samples.

When transferring corrosive material, use a **pipet (pipette) filling attachment**.

Noxious samples can be safely handled with a **pipet (pipette) filling bulb**.

AUTOMATIC TRANSFER PIPET (Automatic Transfer Pipette)

An **automatic transfer pipet (pipette)** is used for rapid and reproducible dispensing of liquids.

Approximate accuracy is + 2% for the **automatic transfer pipet (pipette)**.

An **automatic transfer pipet (pipette)** is used for multiple filling operations.

GRADUATED CYLINDER

This is a 250 ml **graduated cylinder**.

Use the **graduated cylinder** to measure volume when extreme accuracy is not important.

Some **graduated cylinders** have stoppers fitted to the top.

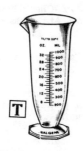

GRADUATE

A **graduate** is less accurate than a **graduated cylinder.**

Measure the approximate volume with the **graduate.**

Many **graduates** are calibrated in both fluid ounces and milliliters.

BEAKER

Some **beakers** are calibrated for approximate volume.

Measure approximately 100 ml with the **beaker.**

A **beaker** cannot accurately measure volume.

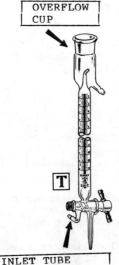

OVERFLOW
CUP

GRADUATED BURET (Graduated Burette)

A **graduated buret (burette)** is used to measure volume.

This **graduated buret (burette)** has an overflow cup and automatically zeros when filled.

Accurately measure the volume with the **graduated buret (burette).**

INLET TUBE
FOR REFILLING

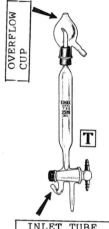

OVERFLOW CUP

AUTOMATIC ZERO PIPET (Automatic Zero Pipette)

Measure 25 ml with the **automatic zero pipet (pipette)**.

The **automatic zero pipet (pipette)** can accurately measure a specific volume.

This **automatic zero pipet (pipette)** has an inlet tube for refilling.

INLET TUBE FOR REFILLING

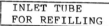

PLUG

STOPCOCK

The **stopcock** controls flow in or out of a **buret (burette)**.

The clip on the **stopcock** holds the plug in place.

This type of **stopcock** creates a Y-type connection.

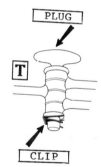

CLIP

STOPCOCK GREASE

Stopcock grease is used to lubricate stopcocks.

Desiccators can be sealed with **stopcock grease**.

Stopcock grease is acidic and alkaline resistant.

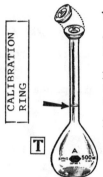

VOLUMETRIC FLASK

This **volumetric flask** contains 500 ml.

Most **volumetric flasks** are calibrated at 20 degrees centigrade.

Measure the volume accurately with the **volumetric flask**.

PUMP

A **pump** is used to transfer chemicals.

An acid can be transferred with a chemical resistant **pump**.

Transfer the solvent with a **pump**.

Measurement of volume is a common laboratory procedure
and many instruments are designed to accomplish this task. A
syringe pipet (pipette) is used to accurately measure
small volumes. A **transfer (or volumetric) pipet
(pipette)** is used to accurately measure large specific
volumes but a **measuring (or graduated) pipet (pipette)**
is necessary for measuring varying volumes. If there is fear
of getting the chemical in one's mouth, a **pipet (pipette)
filling attachment (or bulb)** should be used. An
automatic transfer pipet (pipette) is useful when a
repeated transfer of a specific volume is made. **Graduated
cylinders** are less accurate than a **volumetric** or
measuring pipet (pipette) but more accurate than a
graduate or a **calibrated beaker.** A **graduated buret
(burette)** or an **automatic zero pipet (pipette)** also
accurately measures volume. A **buret (burette)** is fitted
with a **stopcock** to meter the liquid from the **buret
(burette)** and the **stopcock** is lubricated with
stopcock grease. A **volumetric flask** is useful when
accuracy of large specific volumes is needed. **Pumps** are
used in the laboratory to transfer liquids.

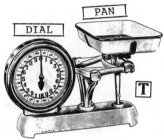

SCALE

Use the **scale** to weigh a sample of medium size.

Some **scales** operate with the aid of a spring.

Scales are usually not as accurate as balances.

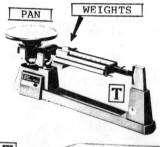

TRIPLE BEAM BALANCE

A **triple beam balance** functions with the aid of a sliding weight on 3 beams.

A sensitivity of 0.1 gram is often obtained with a **triple beam balance.**

Weigh the sample on the **triple beam balance.**

BALANCE

The semi-micro **balance** often has a precision of $\pm$ 0.01 mg.

Weights are added or removed from this **balance** by means of dials.

Readout on this **balance** is digital.

reading 118.47325 g

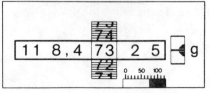

TOP-LOADING BALANCE

This **top-loading balance** is an electronic digital type.

The **top-loading balance** is very easy to use.

Weigh the sample on the **top-loading balance**.

WEIGHT SET

Weight sets are used with some balances.

Do not handle the **weights** with your fingers.

Check this balance with **weight set**.

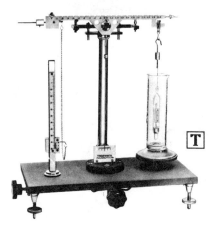

SPECIFIC GRAVITY BALANCE

A **specific gravity balance** has the ability to weigh a sample in a liquid.

Determine the **specific gravity** of the sample with the **balance**.

Lead has a very high **specific gravity** when determined on the **balance**.

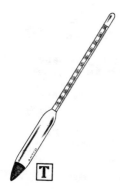

HYDROMETER

A **hydrometer** is used to determine specific gravity.

A **hydrometer** is floated in the liquid.

The specific gravity of the salt solution is determined with the **hydrometer**.

SPECIFIC GRAVITY BOTTLE OR PYCNOMETER

Use the **specific gravity bottle (pycnometer)** to weigh an exact volume of liquid.

Fill the **specific gravity bottle (pycnometer)** with water and then with the sample.

Determine **specific gravity** with the **pycnometer**.

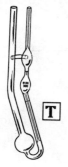

VISCOMETER

Use the **viscometer** to determine viscosity.

Rate of flow is determined by a **viscometer**.

Check the sample with the **viscometer**.

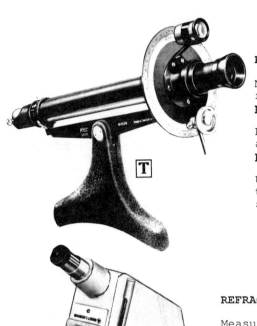

POLARIMETER

Measure the angle of rotation with the **polarimeter.**

Place the optically active sample in the **polarimeter.**

Use the **polarimeter** to check a sugar solution.

REFRACTOMETER

Measure the percentage of sugar with the **refractometer.**

Use the **refractometer** to measure the total dissolved solids.

Percentage alcohol can be determined with a **refractometer.**

HAND REFRACTOMETER

Check alcohol content of a wine sample with a **hand refractometer.**

The **hand refractometer** can be taken to the field for use.

Place the syrup in the **hand refractometer.**

Weighing is an important task performed in the laboratory. Heavy items are weighed on a **scale** and light items are weighed on a **balance**. A **triple beam balance** is used when accuracy is not critical. The electronic digital **top-loading balance** is very easy to use when weighing light objects. **Weight sets** are used on older **balances** and to check the accuracy of **balances**. **Balances** that can weigh the samples suspended in liquid are useful for determining **specific gravity**. The **specific gravity** of a liquid can be determined by a floating **hydrometer**. A **specific gravity bottle** or **pycnometer** allows one to weigh a specific volume of material. The rate of flow of a liquid can be measured with a **viscometer**. A **polarimeter** can be used to measure the rotation of light and a **hand refractometer** can be used to measure light deflection.

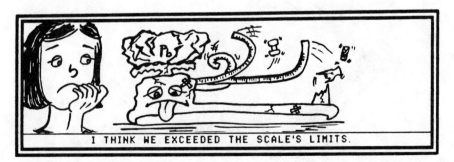

I THINK WE EXCEEDED THE SCALE'S LIMITS.

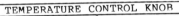

OVEN

This **oven** is a vacuum **oven**.

Other **ovens** are convection or forced air **ovens**.

Dry the sample in the **oven**.

HOT PLATE

A **hot plate** is used for heating containers.

The temperature of most **hot plates** is controllable.

A **hot plate** can be used in place of a stove heating element or gas flame.

INCUBATOR

This is a CO_2 **incubator**.

Place the petri dish in the **incubator**.

The **incubator** maintains biological growing temperatures.

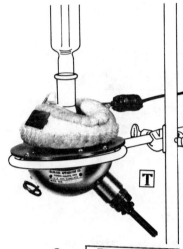

HEATING MANTLE

Place the **heating mantle** around the flask.

Connect the **heating mantle** to a transformer.

The hemispherical **heating mantle** will enclose the flask.

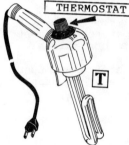

THERMOSTAT

IMMERSION HEATER

Place the **immersion heater** in the water bath.

Keep the liquid temperature constant with the **immersion heater**.

The **immersion heater** is portable.

PLUG

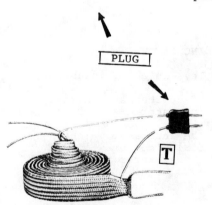

HEATING TAPE

Wrap the **heating tape** around the column.

The flexible flat **heating tape** is useful in heating glassware.

Heating tape is available in many sizes and wattages.

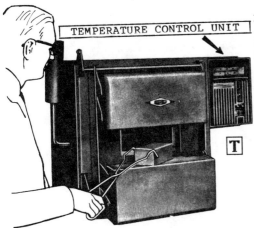

MUFFLE FURNACE

A **muffle furnace** is a high temperature furnace.

Determine the ash content of the sample by placing it in the **muffle furnace.**

 Use tongs when placing samples in the **muffle furnaces.**

CRUCIBLE

A **crucible** is a container that can withstand high temperatures.

A **crucible** may be made from porcelain, graphite, or platinum.

Place the **crucible** in the muffle furnace.

WATER BATH

A **water bath** is used to maintain a constant temperature.

A **water bath** may be electric, gas or steam heated.

Heat the sample in the **water bath.**

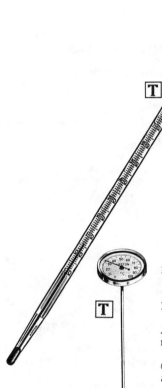

THERMOMETER

Use the **thermometer**
to measure the
temperature.

A **thermometer** may be
calibrated in Celsius
(^{o}C) or Fahrenheit
(^{o}F).

The **thermometer** is
usually filled with
mercury.

DIAL THERMOMETER

The **dial thermometer** contains a
bimetallic sensing element.

A **dial thermometer** is
moderately rugged.

Celsius and Fahrenheit scales are
available on **dial thermometers**.

THERMOMETER (Electrical)

This **thermometer** allows
you to monitor temperature
electronically.

This **thermometer** may be
attached to a recorder to give
a chart of time and
temperature.

Some **thermometers** monitor
temperature at more than one
position with multiple probes.

AUTOCLAVE OR STERILIZER

The **autoclave** or
sterilizer may be electric
(generates steam) or steam
heated.

Sterilize the microbiological
samples in the **autoclave**.

Sterilize at 15 p.s.i. in the
autoclave.

INSTRUMENT STERILIZER

An **instrument sterilizer**
uses hot water to sterilize.

Autoclave temperatures are
higher than **instrument
sterilizers**.

Sterilize the equipment in the
instrument sterilizer.

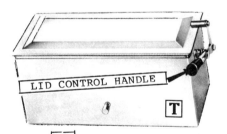

LID CONTROL HANDLE

BURNER

A gas **burner** is used for
heating samples in the laboratory.

Heat the sample in the **burner**
flame.

Adjust the air on the **burner**.

AIR ADJUSTMENT

GAS ADJUSTMENT

ALCOHOL LAMP

The **alcohol lamp** gives you a portable flame.

Extinguish the flame and add the cap when you are moving the **alcohol lamp**.

Adjust the wick on the **alcohol lamp**.

GAS LIGHTER

The **gas lighter** produces a spark for ignition.

Put a new flint in the **gas lighter**.

The flint strikes the rasp to produce the spark in the **gas lighter**.

SAFETY MATCHES

Light the burner with the **safety matches**.

For the **safety match** to be ignited, it must be struck against the box.

A **gas lighter** may be used instead of **safety matches**.

MELTING POINT BATH

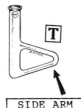

SIDE ARM

Heat the side arm on the **melting point bath.**

Determine the sample melting point with a **melting point bath.**

The **melting point bath** is used to raise the temperature slowly.

MELTING POINT TUBE

Place the sample in a **melting point tube.**

Place the **melting point tube** in the **melting point bath.**

By slowly heating the **melting point tube,** one can determine the melting point of the sample.

FURNACE NOT WORKING ?

In the laboratory, **heating** can be accomplished with
an **oven** or with a **hot plate**. The **incubator** can be
used to maintain a constant growing temperature. Glassware may
be heated with a **heating mantle** or with a **heating tape**. An
immersion heater is used to heat liquid. Samples may
be heated to a very high temperature by placing them in a
crucible which is then placed in a **muffle furnace**. A
water bath is used to maintain a temperature between room
temperature and 100°C. Temperature may be measured with a
mercury thermometer, a **dial thermometer,** or an
electronic thermometer. Sterilization can be accomplished
in a steam **autoclave** or in a hot water **instrument
sterilizer.** Flames may be maintained in the laboratory
with a gas **burner** or with an **alcohol lamp.** Gas flames
may be ignited with a **gas lighter** or with a **safety
match.** The melting point of a sample may be determined by
using a **melting point tube** in a **melting point bath.**

Please handle with care

It is the only one we have !!!

MORTAR AND PESTLE

Mortars and pestles are made of porcelain, glass or metal.

Grind the sample in the **mortar and pestle**.

Mortars and pestles are available in many sizes.

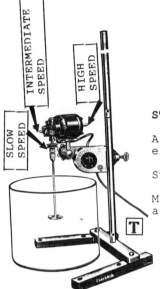

STIRRER

A **stirrer** may be air driven or electrically driven.

Stir the liquid with the **stirrer**.

Mix the powder into the liquid with a **stirrer**.

MAGNETIC STIRRER

A **magnetic stirrer** can be used in a closed vessel.

Some hot plates are equipped with a **magnetic stirrer**.

A rheostat controls the speed of the **magnetic stirrer**.

LARGE SIZE SHAKER

This **large size shaker** can shake large containers.

This **shaker** is of the reciprocating type.

Extract the component from the tissue by using the **large size shaker**.

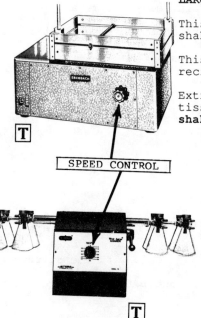

SPEED CONTROL

SHAKER

This is a flask-type **shaker**.

The flask **shaker** gives a snap-wrist motion.

The flask **shaker** produces a swirling effect.

BLENDER

Mince the sample in the **blender**.

The **blender** can be used to homogenize the sample.

Blade speed of the **blender** can be as much as 21,000 r.p.m.

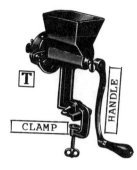

HAND GRINDER (Hand Mincer)

A **hand grinder (mincer)** can be used to grind (mince) dry or wet material.

Grinding plates can be changed on the **hand grinder (mincer)** to change the fineness of grind (mince).

A worm drive moves the sample through the **hand grinder (mincer)**.

MILL

The **mill** is used for reduction in size of analytical samples.

The **mill** uses a shearing action.

Sieves prevent only small particles from passing through the **mill**.

TISSUE GRINDER

The **tissue grinder** turns a grinding pestle.

Bacterial cells and soft tissue can be ground in a **tissue grinder**.

The **tissue grinder** breaks up tissue between ground glass surfaces.

HOMOGENIZER

The sample can be emulsified by using the **homogenizer.**

Disintegration of biological cells can be accomplished by a **homogenizer.**

The **homogenizer** works on a shearing principle.

EMULSION HOMOGENIZER

The **emulsion homogenizer** produces a dispersion from immiscible liquids.

High pressure is produced at the orifice of the **emulsion homogenizer.**

The **emulsion homogenizer** is useful for small quantities.

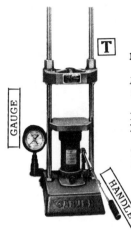

HYDRAULIC PRESS

A **hydraulic press** can be used for separating liquid from solid.

Pellets can be made with a **hydraulic press**.

This **hydraulic press** develops 24,000 p.s.i.

YOUR SAMPLE NOT YOUR THUMB GOES IN THE PRESS !

Mixing and pressing equipment is very useful in the laboratory. The simplest grinding and mixing equipment is a **mortar and pestle**. **Stirrers** may be of the **propeller** type or may be **magnetic** in style. **Shakers** may be **reciprocating** or have a **wrist type** motion. **Blenders** can also be used for grinding and mixing. **Hand grinders (mincers)**, **mills** and **tissue grinders** are all used to break up a sample. **Homogenizers** may use a shearing principle or pressure through an orifice to produce a dispersion. Powder may be formed into a pellet with a **hydraulic press.**

I'LL GET A GOOD MIX OF MY CHEMICALS THIS TIME.

pH METER

The hydrogen ion concentration is measured with a **pH meter**.

Use the **pH meter** to see if the sample is acid.

The pH of the sample is determined with a **pH meter**.

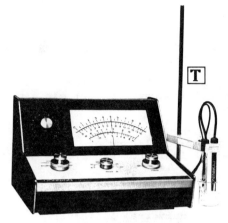

ELECTRODE

REFERENCE

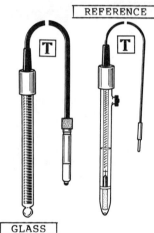

GLASS

Two **electrodes** are attached to the **pH meter**.

A combination **electrode** contains both a glass and a reference electrode in one case.

Keep the **electrodes** clean for accurate pH measurement.

pH BUFFER SOLUTION

Use a **pH buffer solution** to standardize the **pH meter**. •

Buffer solutions are available in many pH values.

Standardize the **pH meter** with a **buffer solution** that has a pH near the pH of the sample to be measured.

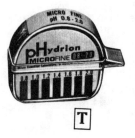

TEST PAPER OR "LITMUS" PAPER OR UNIVERSAL INDICATOR PAPER

Test paper (or litmus or universal indicator paper) can measure pH by a color change.

Check the pH of the sample with the **test paper (or litmus or universal indicator paper)**.

The **pH meter** can more accurately determine sample pH than you can with the **test paper (or litmus or universal indicator paper)**.

VOLT-OHM-AMMETER (AVO-meter)

This **volt-ohm-ammeter (AVO-meter)** can measure voltage, resistance and amperes.

Both a.c. and d.c. voltages can be measured on the **volt-ohm-ammeter (AVO-meter)**.

There are 6 ranges to measure resistance on the **volt-ohm-ammeter (AVO-meter)**.

BAROMETER

Atmospheric pressure is measured with a **barometer**.

The **barometer** on the left is filled with mercury.

The **barometer** on the right is actuated by an airtight metal diaphragm.

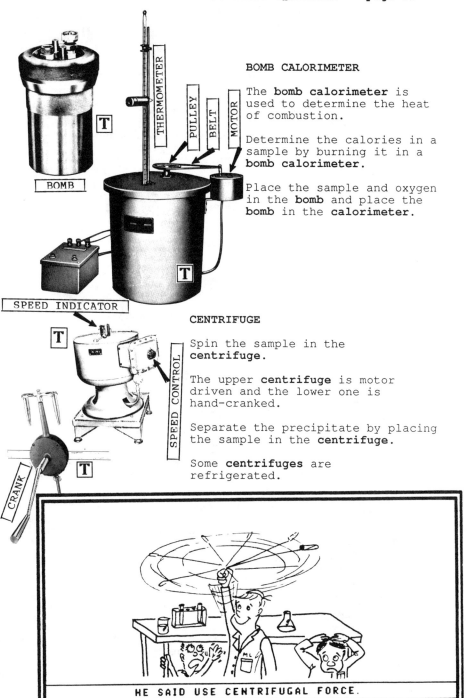

BOMB CALORIMETER

The **bomb calorimeter** is used to determine the heat of combustion.

Determine the calories in a sample by burning it in a **bomb calorimeter**.

Place the sample and oxygen in the **bomb** and place the **bomb** in the **calorimeter**.

CENTRIFUGE

Spin the sample in the **centrifuge**.

The upper **centrifuge** is motor driven and the lower one is hand-cranked.

Separate the precipitate by placing the sample in the **centrifuge**.

Some **centrifuges** are refrigerated.

THERMOMETER

PULLEY

BELT

MOTOR

BOMB

SPEED INDICATOR

SPEED CONTROL

CRANK

HE SAID USE CENTRIFUGAL FORCE.

CENTRIFUGE HEAD (Rotor)

Place the sample tube in the **centrifuge head (rotor).**

Different **centrifuge heads (rotors)** can be placed on the centrifuge.

Different size sample containers may be placed in different **centrifuge heads (rotors).**

CENTRIFUGE TUBE

Place the **centrifuge tube** in the **centrifuge head (rotor).**

Place **centrifuge tubes** on opposite sides of the **centrifuge head (rotor)** to balance the weight during spinning.

Centrifuge tubes may be either glass, plastic or metal.

CARRIERS

Carriers fit on **centrifuge heads (rotors)** and hold sample containers.

With this **head** and **carriers,** the centrifuge will hold 168 small **tubes.**

Other **carriers** will fit this **head** enabling it to hold 6 large **centrifuge** bottles.

SPECTROPHOTOMETER

Measure the optical density of the sample with the **spectrophotometer**.

Set the desired wavelength on the **spectrophotometer** prior to measuring the sample.

Concentration of samples will influence the **spectrophotometer** reading.

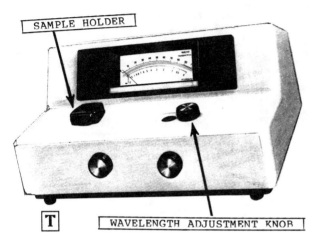

SAMPLE HOLDER

T

WAVELENGTH ADJUSTMENT KNOB

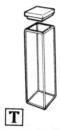

T

SPECTRO CELL

Spectro cells may be round, square or rectangular.

Do not scratch the **spectro cells**.

Use matched **spectro cells** for the blank and the sample.

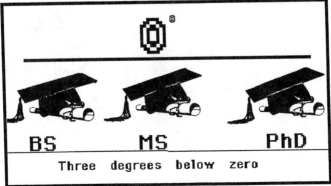

BS   MS   PhD

Three degrees below zero

FLUOROMETER (Fluorimeter)

The **Fluorometer (fluorimeter)** is used for quantitation of trace materials.

The **fluorometer (fluorimeter)** has controllable emission and excitation wavelengths.

Measure the vitamin content with the **fluorometer (fluorimeter)**.

T

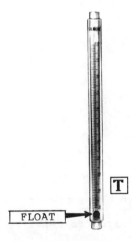

FLOAT

FLOWMETER

A **flowmeter** can measure rate of flow of a gas or of a liquid.

T

The height of the float in the **flowmeter** indicates flow rate.

Flowmeters are available in a variety of flow rate ranges.

MEASURING AIR FLOW ?

The laboratory is equipped with many instruments for measuring and the **pH meter** with its glass and reference **electrodes** is an example. The **pH meter** should be standardized with **buffer solutions.** The hydrogen ion concentration can also be approximated with **pH test paper (litmus or universal indicator paper).** Many electrical measurements can be made with **volt-ohm-ammeter (AVO-meter).** A **barometer** is used to measure barometric pressure. Calories can be measured with a **bomb calorimeter.** A **centrifuge** is used to separate materials based on their density and is necessary in many gravimetric analyses. **Centrifuge heads (rotors), carriers** and **tubes** are interchangeable, giving the **centrifuge** a great deal of flexibility. A **spectrophometer** is a very useful laboratory measuring device, and many types of **spectro cells** can be used. A **fluorometer (fluorimeter)** is often useful in determining concentrations. A **flowmeter** can measure flow rate of a gas or of a liquid.

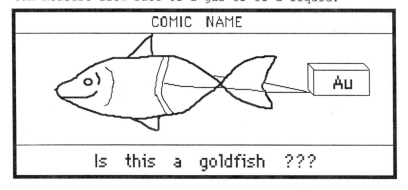

COMIC NAME

Is this a goldfish ???

GEL COLUMN ELECTROPHORESIS

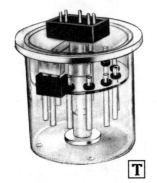

Use the **gel column electrophoresis** to separate the protein mixture.

Connect the electrical power supply to the **gel column electrophoresis** cell.

Add the appropriate buffers to the **gel column electrophoresis** cell.

SCANNING DENSITOMETER

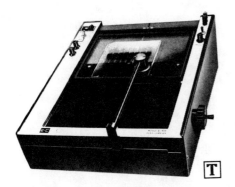

Use the **scanning densitometer** to measure the absorbance of the gel column.

Quantitative evaluation of the electrophoretic patterns is possible with a **scanning densitometer**.

Adjust the **scanning** wavelength on the **densitometer**.

DIALYZER

Use the **dialyzer** for concentration or purification of biological solutions.

Place the tubing containing the sample in the **dialyzer**.

This **dialyzer** rotates the sacs in the bath to stir both the bath and sac contents.

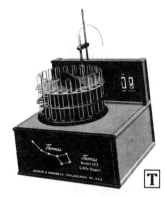

AUTOMATIC SAMPLER

The **automatic sampler** allows for unattended transfer of serial samples.

The electrode can be automatically lowered into serialized samples for measurement with an **automatic sampler.**

Some **automatic samplers** can be used to fill tubes.

SIEVE

A **sieve** separates samples by particle size.

Size of opening in a **sieve** is often designated as "mesh size".

Sieve size is sometimes specified by National organizations.

SEPARATORY FUNNEL (Separating Funnel)

A **separatory (separating) funnel** is used to separate immiscible liquids into separate phases (or layers).

Drain the lower liquid (phase) from the **separatory (separating) funnel.**

Hold the stopper and stopcock when shaking the **separatory (separating) funnel.**

CHROMATOGRAPHY COLUMN

Separate the sample with a
chromatography column.

Fill the **chromatography column**
with solvent and packing material
and then add the sample to the top
of the column and run more solvent
through it.

Collect the effluent as it comes
from the **chromatography column**.

THIN LAYER CHROMATOGRAPHY (TLC)

Thin layer chromatography is
usually accomplished on glass
plates.

Separate the sample on the **thin
layer chromatography** plate.

What solvent should I use for
separation of the sample on the
thin layer chromatography
plate?

CHROMATOGRAPHY CABINET

Place the filter paper in the
chromatography cabinet.

Separate the samples by placing
them in the **chromatography
cabinet**.

The **chromatography cabinet** is
solvent tight.

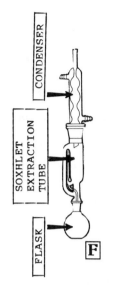

CONDENSER

SOXHLET EXTRACTION TUBE

FLASK

F

SOXHLET EXTRACTOR

Extract the fat from the sample with a **Soxhlet extractor.**

Place the **Soxhlet extractor** on a hot plate.

Pass cold water through the condenser.

Ether is the solvent most often used in a **Soxhlet extractor.**

EXTRACTION THIMBLE

T

Place the sample in the **extraction thimble.**

The **extraction thimble** is placed in the extraction tube of the **Soxhlet extractor.**

This **extraction thimble** is made of cellulose.

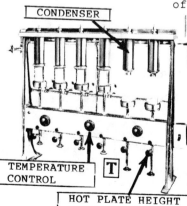

CONDENSER

TEMPERATURE CONTROL

T

HOT PLATE HEIGHT ADJUSTMENT

GOLDFISCH EXTRACTOR

Determine the fat in the food sample on the **Goldfisch extractor.**

The **Goldfisch extractor** normally requires less time than the **Soxhlet extractor.**

Slide the heater up under the beaker on the **Goldfisch extractor.**

BABCOCK BOTTLE

Measure the fat content of the
sample with the **Babcock bottle.**

Different styles of **Babcock
bottles** are used for different
types of samples.

Fat content of dairy products is
often measured with **Babcock
bottles.**

DIVIDER

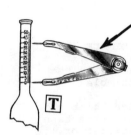

Measure the fat column with the
divider.

A **divider** measures linear
distance.

Place the points of the **divider**
at the top and bottom of the fat
column.

BABCOCK CENTRIFUGE

BRAKE

Place the **Babcock bottle** in the
centrifuge.

By using the **Babcock centrifuge,**
one obtains better fat separation.

This **Babcock centrifuge** is
temperature controlled.

ALUMINUM FOIL DISH (Aluminium Foil Dish)

A disposable **aluminum (aluminium) foil dish** can be used for drying samples.

Place the **aluminum (aluminium) foil dish** in the drying oven.

Dry the **aluminum (aluminium) foil dish** prior to use.

DESICCATOR

Allow the **aluminum foil dish** to cool in the **desiccator**.

Some **desiccators** are equipped with a vacuum top.

Place the drying agent in the bottom of the **desiccator**.

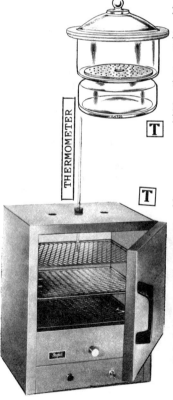

DRYING OVEN

Determine the moisture content of the sample by placing it in the **drying oven**.

Regulate the **drying oven** temperature at 100 degrees centigrade.

Be sure all flammable volatile chemicals have evaporated from the sample prior to placing it in the **drying oven**.

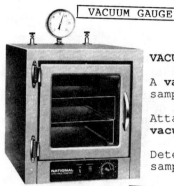

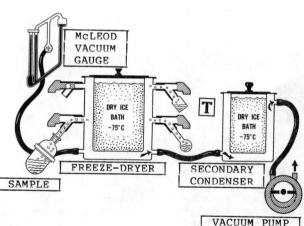

VACUUM OVEN

A **vacuum oven** can accelerate sample drying.

Attach a vacuum pump to the **vacuum oven**.

Determine the moisture level of the sample by using the **vacuum oven**.

MOISTURE BALANCE

Determine sample moisture on the **moisture balance**.

This **moisture balance** has infrared heating.

Adjust the timer on the **moisture balance**.

FREEZE DRYING

Freeze drying is a method of removing water from solid samples.

Moisture level can be determined by **freeze drying**.

Freeze drying is a method of preservation.

Freeze drying of microorganisms is useful when storage is desirable.

KJELDAHL FLASK

To determine protein, place the sample in the **Kjeldahl flask.**

Kjeldahl flasks are available in several sizes.

Add acid to the **Kjeldahl flask.**

KJELDAHL DIGESTION UNIT

Place the **Kjeldahl flask** on the **Kjeldahl digestion unit.**

The **Kjeldahl digestion unit** heats the flask with gas or with an electric hot plate.

Fumes from the **Kjeldahl digestion unit** must be vented through a hood or down the drain.

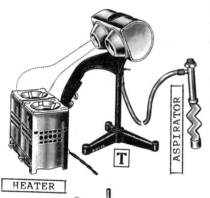

HEATER

ASPIRATOR

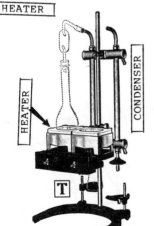

HEATER

CONDENSER

KJELDAHL DISTILLING UNIT

After digestion, place the flask on the **Kjeldahl distilling unit.**

Pass cold water through the condenser of the **Kjeldahl distilling unit.**

Heat the **Kjeldahl flask** on the **Kjeldahl distilling unit.**

RECORDER

The **recorder** can be attached to many types of measurement equipment.

A **recorder** attached to the appropriate measurement equipment can be used to determine concentration in a sample.

Speed of the **recorder** and deflection sensitivity can be adjusted.

T

I RECORDED U.S.A. ???

Separation equipment performs an important function in the laboratory. **Gel column electrophoresis** separates samples by electrical charges and quantitative information can be obtained by placing the columns on a **scanning densitometer.** The **dialyzer** and **sieve** separate samples by size. The **automatic sampler** is very useful in automating many of the separation instruments. Partitioning of a sample between two liquids can be accomplished by using a **separatory (separating) funnel.** Many types of **chromatography** such as **column, thin layer, paper** or **gas,** can separate a complex sample into components. Fat can be separated by placing a sample in an **extraction thimble** and solvent-extracting the sample in a **Soxhlet extractor** or a **Goldfisch extractor.** Fat can also be determined by placing the sample in a **Babcock bottle;** and, after digestion--placing the bottle in a **Babcock centrifuge.** The percentage of fat is determined with the aid of a **divider.** Sample moisture can be determined by placing the sample in an **aluminum (aluminium) foil dish** and placing this in a **drying oven** or in a **vacuum oven.** After drying, the sample is cooled in a **desiccator.** Moisture can also be determined with a heat lamp and a **moisture balance** or by **freeze drying** the sample. For protein determination, place the sample in the **Kjeldahl flask** and place the flask on the **digestion unit.** After digestion, distill the nitrogen on the **distilling unit.** Protein is then determined by titration. A **recorder** is useful in quantifying the results from many of the separation instruments.

FUNNEL

Funnels are available in glass, plastic, porcelain, and metal.

Stem length, angle and diameter of body vary in different **funnels.**

Filter the sample through the **funnel.**

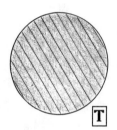

FILTER PAPER

Filter paper may be flat or folded.

Number of the **filter paper** indicates filtering speed.

Place the **filter paper** in the **funnel.**

GLASS WOOL

Glass wool can be used for a filtering material.

Fiber diameter of **glass wool** is normally between 0.005 to 0.008 mm.

Filter the sample through **glass wool.**

FILTERING FLASK (Filter Flask)

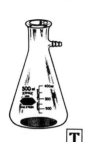

Attach a vacuum hose to the side arm of the **filtering (filter) flask.**

Vacuum filtering can be accomplished with the aid of a **filter flask.**

The filtering rate can be accelerated by using a vacuum and a **filtering (filter) flask.**

BUCHNER FILTER FUNNEL

Place the **Buchner filter funnel** on a **filter flask.**

The **Buchner filter funnel** is used for vacuum filtering.

Buchner filter funnels are usually made of porcelain but sometimes plastic is used.

CRUCIBLE

Crucibles are made of porcelain, glass, nickel or iron.

Some **crucibles** have small holes in the bottom and others have fritted glass bottoms.

Filter the sample through the **crucible.**

CRUCIBLE HOLDER

Place the **crucible** in the **crucible holder**.

Crucible holders can be placed on a **filter flask** or on a standard **funnel**.

Most **crucible holders** are made of rubber.

IMPINGER

A dust sample collector is called an **impinger**.

An **impinger** collects particles in the liquid-filled receiver.

Collect dust samples from the air with an **impinger**.

HE SAID USE A PAPER FILTER.

Filtering separates solids from liquids. A
funnel is used with filter paper or glass wool.
To accelerate filtering, a vacuum is often used with a
filtering (filter) flask and a Buchner filter funnel.
A crucible and a crucible holder can also be used.   To
collect dust from air, an impinger is used.

THAT'S A STRONG FILTER.

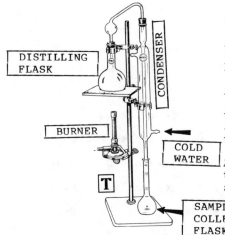

STILL

A **still** separates chemicals based on their boiling points.

A sample in a **still** is first changed to a volatile state by heating and fractions of the vapor are then condensed.

A **still** can be used to concentrate the sample.

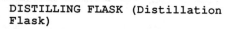

DISTILLING FLASK (Distillation Flask)

The sample is boiled in the **distilling (distillation) flask**.

Place a thermometer in the top of the **distilling (distillation) flask**.

The **distilling (distillation) flask** has a side arm.

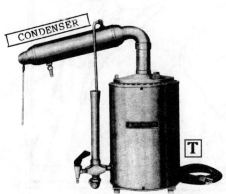

WATER STILL

A **water still** is used to produce **distilled water**.

A **water still** removes water from the minerals in tap water.

A **water still** may be electrical, gas or steam heated.

DEMINERALIZER

A **demineralizer** can also remove minerals from water.

Demineralized water is often used in place of **distilled water**.

Most **demineralizers** use a resin type bed.

T

FLASH EVAPORATOR

A **flash evaporator** works like a **still**.

T

Remove the solvent with a **flash evaporator**.

The **flash evaporator** contains an evaporating bath and a condensing bath.

DISTILLATION DOES COME IN HANDY

A **distillation unit** or **still** will separate
chemicals with different boiling points. The major components
of a **still** are a heat source, a **distilling
(distillation) flask** and a condenser. **Water stills**
are frequently found in a laboratory to produce **distilled
water** and **demineralized water** is produced by passing
the water through a resin-type bed which traps the minerals.
A **flash evaporator** uses the distillation principle to
remove a solvent from the sample.

THEY SAY AN ICEBERG IS DEMINERALED WATER.

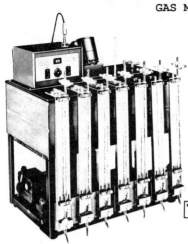

RESPIROMETER

Gas exchange studies can be performed on a **respirometer**.

A **respirometer** is often called a Warburg unit.

Attach the manometer to the **respirometer**.

THE CANARY CAN ALSO MEASURE GAS.

A **respirometer** permits precise measurements of gas evolved or absorbed in cell respiration.

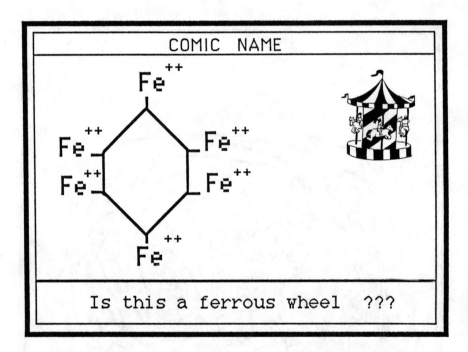

COMIC NAME

Is this a ferrous wheel ???

ANIMAL CAGE

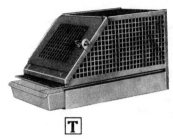

Place the mouse in the **animal cage.**

Some **animal cages** are made of metal and others are made of plastic.

Be sure the **animal cage** is of the appropriate size for the animal.

DRINKING TUBE

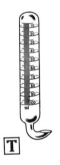

Attach the **drinking tube** to the **animal cage.**

Keep the **drinking tube** filled with clean water.

This **drinking tube** is graduated.

FEEDING DISH

Place the animal food in the **feeding dish.**

Clean the **feeding dish** frequently.

The food level in the **feeding dish** should be checked daily.

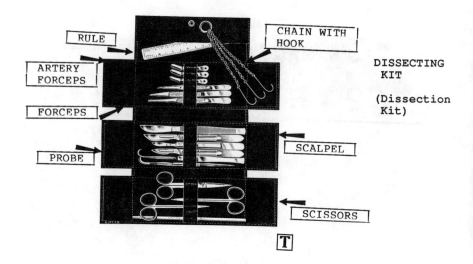

RULE

CHAIN WITH HOOK

ARTERY FORCEPS

FORCEPS

PROBE

DISSECTING KIT

(Dissection Kit)

SCALPEL

SCISSORS

This **dissecting kit** contains a **rule, forceps, artery forceps, probe, scissors, scalpels** and a **chain with hooks.**

Use the **dissecting kit** to separate the tissue.

Get the **scalpel** from the **dissecting kit.**

ANIMAL OPERATING TABLE

A small **animal operating table** without legs is called an animal board.

Perform the surgery on the anesthetised animal after placing it on the **operating table.**

Sanitize the **animal operating table** before the operation.

Place the rabbit in the **animal cage** and fill the
drinking tube and the **feeding dish**. Prepare for
surgery by cleaning the **animal operating table** and
checking the **dissecting (dissection) kit**. Be sure that
there are sufficient **forceps** and **probes** and that the
scissors and **scalpels** are sharp.

I DON'T THINK THE ANIMAL FITS THE CAGE.

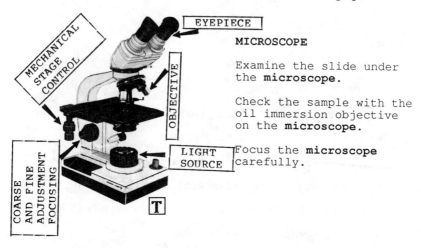

EYEPIECE

MECHANICAL STAGE CONTROL

OBJECTIVE

COARSE AND FINE ADJUSTMENT FOCUSING

LIGHT SOURCE

MICROSCOPE

Examine the slide under the **microscope**.

Check the sample with the oil immersion objective on the **microscope**.

Focus the **microscope** carefully.

SLIDE

Place the microorganisms on the **slide**.

Identify the **slide** on the marking surface.

View the **slide** on the **microscope**.

MAGNIFIER

A **magnifier** will change the size 2 to 5 times.

Focus the **magnifier** by moving it in relation to the object being viewed.

Count the bacterial colonies with the **magnifier**.

LENS TISSUE

Use only **lens tissue** to clean optical lens.

Clean the objective on the microscope with the **lens tissue.**

A **lens tissue** is soft, free from impurities and does not easily collect dust.

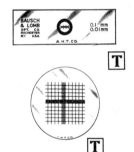

MICROMETER

A **micrometer** is used when a measurement with a microscope is needed.

Some **micrometers** are placed on the stage (upper picture) and some are placed in the eyepiece (lower picture).

Measure the mold filament length with the **micrometer.**

COUNTING CHAMBER

Place the blood in the **counting chamber** to count the cells.

Count the bacteria with the aid of the **counting chamber.**

Check the food for mold mycelia with the **counting chamber.**

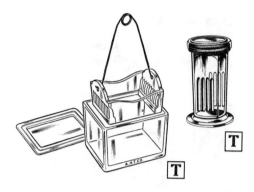

STAINING DISH OR JAR

Stain the tissue on the slides in the **staining dish** or **jar**.

The **staining dish (or jar)** may be used to aid in identifying the bacteria.

A **staining dish** may be used for staining, differentiation, dehydration and clearing of embedded sections.

TISSUE PROCESSING UNIT

The **tissue processing unit** is used for automatically fixing, dehydrating, and infiltrating tissue specimens.

The **tissue processing unit** may be programmed through 12 stations.

A paraffin bath is also located in this **tissue processing unit**.

STATIONS OR BATHS

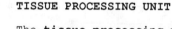

CRANK

MICROTOME

Use the **microtome** to thinly slice the tissue.

The object clamp moves up and down on the **microtome**.

Cutting angle of the knife is adjustable on the **microtome**.

SAMPLE

KNIFE

Microscopic examination is often useful in the laboratory. A **microscope** is used when large magnifications are needed and a **magnifier** is used for small magnification. Biological tissues or microorganisms are often mounted on a **slide** for viewing. Optical lens are cleaned with the aid of **lens tissues**. A **micrometer** is used to measure small distances and a **counting chamber** is used to count small objects. Biological tissues or microorganisms are stained in a **staining dish** or **jar** or in a **tissue processing unit**. This tissue is sliced on a **microtome** prior to viewing.

ARE YOU SURE THIS BACTERIA IS FRIENDLY ?

CULTURE MEDIA

Microbiology **culture media** are usually dehydrated.

Some **culture media** are selective and will only permit growth of certain types of microorganisms.

If the **culture media** contain agar, they will gel when cooled.

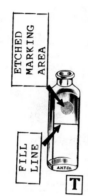

ETCHED MARKING AREA

FILL LINE

DILUTION BOTTLE

Dilute the microorganisms with the aid of a **dilution bottle.**

The **dilution bottle** usually has a capacity of 160 ml.

A rubber stopper with an elongated bottom is used to cap the **dilution bottle.**

PETRI DISH

Petri dishes are usually plastic or glass.

Grow the microorganisms in the **petri dish.**

Pour the culture media into the **petri dish.**

PETRI DISH STERILIZING BOX

Place the **petri dish** in the rack and then put the rack in the **sterilizing box.**

Place the **sterilizing box** in the autoclave.

Most **sterilizing boxes** will hold 12 to 15 petri dishes.

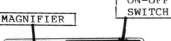

QUEBEC COLONY COUNTER

Place the **petri dish** on the **Quebec colony counter.**

The **Quebec colony counter** is used to count bacteria colonies on a **petri dish.**

The **Quebec colony counter** illuminates and magnifies the **petri dish.**

COUNTER

Use the **counter** to tally the number of microorganisms.

This is a hand **counter** and there are also table models.

The **counters** contain a zero reset knob.

INOCULATING NEEDLE OR LOOP

Use the **inoculating needle** or **loop** to transfer the microorganisms.

Sterilize the **inoculating needle** (or **loop)** in a flame.

The **inoculating loop** will transfer a specific volume of liquid.

HOLDER

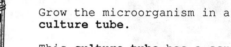

CULTURE TUBE

Grow the microorganism in a **culture tube**.

This **culture tube** has a screw cap.

Transfer the microorganisms to the **culture tube** with an **inoculating needle**.

THIS COLONY LOOKS FAMILIAR.

The **microbiology** area is often used in evaluating
biological spoilage and disease-producing agents.
Microorganisms are diluted with the aid of **dilution**
bottles and placed in a **petri dish**.   Then, **culture**
media is added to the **petri dish**.   Prior to placing
the microorganisms in the **petri dish**, the **dishes**
should be sterilized in a **sterilizing box**.   After the
microorganisms have grown into colonies, they are counted on a
Quebec colony counter and tallied on a hand-**counter**.
Microorganisms can be transferred with an **inoculating**
needle or **loop** and are often added to a **culture**
tube.

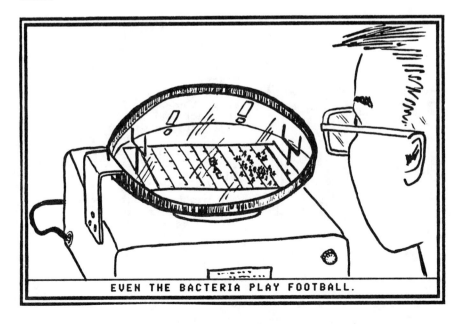

EVEN THE BACTERIA PLAY FOOTBALL.

GLASSWARE WASHER

Use the **glassware washer** after the experiment is finished.

Place the beakers and the detergent in the **glassware washer.**

The last rinse cycle in the **glassware washer** uses distilled water.

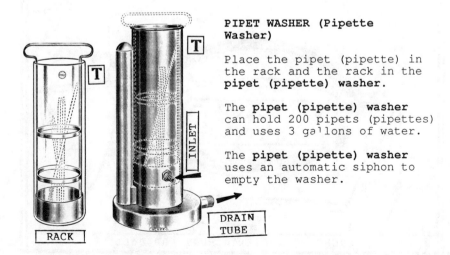

PIPET WASHER (Pipette Washer)

Place the pipet (pipette) in the rack and the rack in the **pipet (pipette) washer.**

The **pipet (pipette) washer** can hold 200 pipets (pipettes) and uses 3 gallons of water.

The **pipet (pipette) washer** uses an automatic siphon to empty the washer.

RACK

DRAIN TUBE

INLET

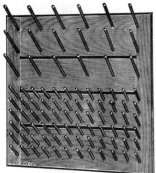

DRAINING RACK

Place the glassware on the **draining rack**.

The **draining rack** may be used for drying equipment.

A **draining rack** can hold glassware in an inverted position.

T

WOULD MAKE A GREAT RING TOSS TARGET !

Use the **glassware washer** to clean the flasks but place the pipets (pipettes) in the **pipet (pipette) washer**. After washing the beakers in the sink and rinsing them in distilled water, place them on the **draining rack**.

THIS WATER IS SURE HARD.

- A -

Acetic Acid 42, 44
Acid 42, 44
ACS 43
Adjustable Wrench 34, 36
Adjusting Knob 34
Air 12, 14, 20, 22
Air Adjustment 77
Air Filter 20
Alcohol 42, 44
Alcohol Lamp 78, 80
Aluminium Foil Dish 99, 103
Aluminum Foil Dish 99, 103
Ammeter 88, 93
Ammonium Hydroxide 42, 44
Ampere Meter 16
Animal Cage 113, 115
Animal Operating Table 114, 115
Apron 27, 32
Artery Forceps 114
Asbestos Gloves 28, 32
Aspirator 21, 22, 101
Atom 45
Autoclave 77, 80
Automatic Sampler 95, 103
Automatic Transfer Pipet 63, 67
Automatic Transfer Pipette 63, 67
Automatic Zero Pipet 65, 67
Automatic Zero Pipette 65, 67
Avoid Contact With Skin 3, 5
AVO-Meter 88, 93

- B -

Babcock Bottle 98, 103
Babcock Centrifuge 98, 103
Back 9
Bag 27, 32
Bail 56
Balance 68, 72, 100, 103
Ball and Socket Joint 51, 55
Ball Point Pen 37, 39
Barometer 88, 93
Base 44
Base Unit 6, 8, 11
Basin 11
Bath 75, 79, 80, 118
Battery 16, 17
Battery Charger 16, 17

Battery Connections   16
Beaker 56, 59, 64, 67
Belt 83, 89
Biohazard 2, 5
Bit 35, 36
Blanket 1, 5
Blender 82, 86
Bomb   89
Bomb Calorimeter 89, 93
Bond   45
Boss Heads 47, 48
Bottle 28, 32, 56, 59,
                                                   70, 72, 98, 103,
                                                   120, 123
Box 24, 25, 78, 121,
                                                   123
Brake   98
Bristle   29
Brush 29, 32
Buchner Filter Funnel 105, 107
Bucket 56, 59
Buffer 87, 93
Bulb 29, 32, 63, 67
Buret 64, 67
Buret Clamp 47, 49
Buret Stand   46
Burette 49, 64, 65, 67
Burette Clamp 47, 49
Burette Stand   46
Burner 77, 80, 108

- C -

Cabinet 6, 7, 8, 96
Cage 113, 115
Calcium Hydroxide 42, 44
Calculator 38, 39
Calibration Ring   66
Calorimeter 89, 93
Ca(OH)$_2$ 42
Cap 18, 29, 57, 59,
                                                   78, 122
Capillary Tubing 50, 55
Carbon Dioxide 23, 24, 25
Carriers 90, 93
Cart 9, 10
Case 7, 8
Catalog Number   43
Caustic 3, 5
Caution 2, 5
Cell 91, 93
Centrifuge 89, 93, 98, 103

Centrifuge Head 90, 93
Centrifuge Rotor 90, 93
Centrifuge Tube 90, 93
Certified 43
Chain 114
Chair 9
Charger 16, 17
Chart 40, 41
CH$_3$CH$_2$OH 42
CH$_3$COOH 42
Checknut 48
Chromatography 96, 103
Chromatography Cabinet 96
Chromatography Column 96, 103
Clamp 18, 19, 47, 48,
                                            49, 51, 55, 60,
                                            61, 83
Clamp Holder 47, 49
Clamp With Holder 48, 49
Cleaning Compound 29, 32
Clip 37, 65
Clock 30, 32
Cold Storage 23, 25
Cold Water 12, 14, 52, 108
Collection Flask 108
Colony Counter 121, 123
Column 96, 103
Concentric Rings 75
Condenser 52, 55, 97, 100,
                                            101, 108
Conical Flask 56, 59
Containers and Closures 56, 59
Cork 58, 59
Cork Borer 58, 59
Corrosive 3, 5
Counter 121, 123
Counting Chamber 117, 119
Coarse Adjustment 116
Crank 89, 118
Crescent Wrench 34, 36
Crown 31
Crucible 75, 80, 105, 107
Crucible Holder 106, 107
Culture Media 120, 123
Culture Tube 122, 123
Cupboard 3, 5, 6, 8
Cutter 37, 54, 55
CW 12
Cylinder 18, 19, 63, 67
Cylinder Coupling 18
Cylinder Gas 18

- D -

Danger 3, 5
Demineralizer 109, 110
Densitometer 94, 103
Desiccator 99, 103
Desk 7, 8
Dewar Flask 24, 25
Dial 2, 68, 76, 80
Dial Thermometer 76, 80
Dialyzer 94, 103
Dilution Bottle 120, 123
Dish 27, 32, 99, 103,
                                               113, 115, 118,
                                               119, 120, 121,
                                               123
Disposable Gloves 28, 32
Dissecting Kit 114, 115
Dissection Kit 114, 115
Distillation 108, 110
Distillation Flask 108, 110
Distilled Water 12, 14
Distilling Flask 108, 110
Distribution Board   15
Distribution Box 15, 17
Divider 98, 103
Door 6, 23
Draining Rack 125, 126
Drain Tube   124
Drawer   6
Drill Bit 35, 36
Drinking Tube 113, 115
Dropping Bottle 29, 32
Dry Ice Bath   100
Dry Ice Box 24, 25
Dry Ice Storage Container 23, 25
Drying Oven 99, 103
DW   12

- E -

Electrical 15, 17
Electrical Outlet 15, 17
Electrical Socket 15, 17
Electrode 87, 93
Electrophoresis 94, 103
Emulsion Homogenizer 84, 86
Erlenmeyer Flask 56, 59
Etched Marking Area   120
Evaporating Dish 27, 32
Evaporator 109, 110
Explosive 2, 5

Extraction Thimble 97, 103
Extraction Tube   97
Extractor 97, 103
Eyepiece   116
Eye Wash 2, 5

- **F** -

Faucet 11, 14
Feeding Dish 113, 115
Felt Tip Marking Pen 38, 39
File 35, 36
Filing Cabinet 7, 8
Fill Line   120
Filter Flask 105, 107
Filtering 104, 105, 107
Filtering Flask 105, 107
Filter Paper 104, 107
Fine Adjustment   116
Fine Nose Pliers   33
Finger   28
Fire Blanket 1, 5
Fire Extinguisher 1, 5
First Aid Kit 1, 5
Fixtures   11
Flammable 3, 5
Flash Evaporator 109, 110
Flask 24, 25, 56, 57,
                                             59, 66, 67, 97,
                                             101, 103, 105,
                                             107, 108, 110
Float   92
Floor Case 7, 8
Florence Flask 57, 59
Flow Control   13
Flow Meter 92, 93
Fluorometer 92, 93
Fluorimeter 92, 93
Foil 27, 32, 99, 103
Foil Cap   27
Foil Dish   27
Forceps   114
Frame 46, 49
Freeze Drying 100, 103
Freezer 23, 25
Fume Cupboard 6, 8
Fume Hood 6, 8
Funnel 95, 103, 104,
                                             105, 107
Furnace 75, 80

Furniture 6, 9
Fuse   15

- G -

Gas 13, 14, 18, 19,
                                                  78, 80
Gas Adjustment   77
Gas Cylinder   18
Gas Cylinder Clamp 18, 19
Gas Lighter 78, 80
Gas Measurement 111, 112
Gauge 18, 20, 21, 85
Gel Column Electrophoresis 94, 103
Glass 50, 55
Glass Electrode   87
Glass Rod 50, 55
Glass Stopcock 52, 55
Glass Stopper 57, 59
Glass Tubing 50, 55
Glass Tubing Cutter 54, 55
Glassware Washer 124, 126
Glass Wool 104, 107
Gloves 28, 32
Goggles 3, 4, 5
Goldfisch Extractor 97, 103
Graduate 64, 67
Graduated Buret 64, 67
Graduated Burette 64, 67
Graduated Cylinder 63, 67
Graduated Pipet 62, 67
Graduated Pipette 62, 67
Grease 65, 67
Grinder 83, 86

- H -

Hammer 34, 36
Hand 30, 31
Hand Grinder 83, 86
Hand Mincer 83, 86
Hand Laboratory Equipment 26, 32
Handle 7, 21, 26, 29,
                                                  34, 56, 83, 85
Hand Pump 21, 22
Hand Pump (Floor Model) 21, 22
Hand Refractometer 71, 72
HCl   42
Heater   101
Heating 73, 80
Heating Mantle 74, 80
Heating Tape 74, 80

Hex Wrench Set 34, 36
HNO_3   42
Holder   122
Hole   57
Homogenizer 84, 86
Hood 3, 5, 6, 8
Hook   114
Hose Connection   18
Hot 3, 5
Hot Plate 73, 80
Hot Plate Height Adjustment   97
Hot Water 12, 14
H_2SO_4   42
HW   12
Hydraulic Press 85, 86
Hydrochloric Acid 42, 44
Hydrometer 70, 72
Hydroxide 42, 44

- I -

Immersion Heater 74, 80
Impinger 106, 107
Incubator 73, 80
Inlet   20
Inlet Tube 64, 65, 124
Inoculating Loop 122, 123
Inoculating Needle 122, 123
Instrument Sterilizer 77, 80
Intake Valve   13
Interval Timer 30, 32

- J -

Jar 118, 119
Joint 51, 55

- K -

Keep In Refrigerator 3, 5
Kjeldahl Digestion Unit 101, 103
Kjeldahl Distilling Unit 101, 103
Kjeldahl Flask 101, 103
Knife   118
KOH   42

- L -

Label 2, 5, 37, 39,
                                           43, 44
Label Maker 37, 39
Laboratory Cart 9, 10
Laboratory Chair   9
Laboratory Furniture   6
Laboratory Hand Tools   33
Laboratory Stool 9, 10
Laboratory Trolley 9, 10
Ladder 9, 10
Lamp 78, 80
Large Size Shaker   82
Leg   7
Lens Tissue 117, 119
Lever 2, 30
Lid Control Handle   77
Lighter 78, 80
Light Source   116
Lip   56
Liquid Volume Measuring Equipment 62
Litmus Paper 88, 93
Lock Release   7
Loop 122, 123
Lot Analysis   43
Lot Number   43

- M -

Magnetic Stirrer 81, 86
Magnifier 116, 119, 121
Mantle 74, 80
Matches 78, 80
McLeod Vacuum Gauge   100
Measurement Equipment 87, 93
Measuring Pipet 62, 67
Measuring Pipette 62, 67
Mechanical Stage Control   116
Media 120, 123
Melting Point Bath 79, 80
Melting Point Tube 79, 80
Metal File 35, 36
Metric   43
Microbiology 120, 123
Micrometer 117, 119
Microscope 116, 119
Microscopic Examination 116, 119
Microtome 118, 119

Mill 83, 86
Mincer 83, 86
Mixer Tap 11, 14
Mixing and Pressing Equipment 81, 86
Mixing Faucet 11, 14
Model 45
Moisture Balance 100, 103
Molecular Models 45
Mortar and Pestle 81, 86
Motor 20, 83, 89
Muffle Furnace 75, 80

- N -

NaOH 42
Needle Nose Pliers 33
Needle Valve 13, 14, 18, 20
NH$_4$OH 42
Nipple 21
NIT 13
Nitric Acid 42, 44
Nitrogen (N$_2$) 13, 14
Notebook 26, 32
Nut 52

- O -

Objective 116
Office Hand Tools 37, 39
Ohm 88, 93
Oil Level 20
Oil Trap 20
On-Off Switch 15, 121
Open-End Wrench 34, 36
Operating 114, 115
O-Ring Joint 51, 55
OSHA 43
Outlet 15, 17, 20
Outlet Box 15
Outlet Strip 15, 17
Out of Order 3, 5
Oven 73, 80, 99, 100, 103
Overflow 11, 13
Overflow Cup 64, 65
OXY 13
Oxygen (O$_2$) 13, 14

- P -

Pail 56, 59
Pan 68, 69
Pen 37, 39
Periodic Chart 40, 41
Periodic Table 40, 41
Pestle 81, 86
Petri Dish 120, 123
Petri Dish Sterilizing Box 121, 123
pH 87, 88, 93
pH Buffer Solution 87, 93
pH Meter 87, 93
Phone   2
Pilot Light   15
Pinch Clamp 51, 55
Pipet 29, 62, 63, 65,
                                                     67, 126
Pipet Filling Attachment or Bulb 63, 67
Pipette 29, 32, 62, 63,
                                                     65, 67, 126
Pipette Filling Attachment 63, 67
Pipette Filling Bulb 63, 67
Pipette Washer 124, 126
Pipet Washer 124, 126
Plastic Stopper   57
Plastic Tubing 60, 61
Pliers 33, 36
Plug 15, 16, 52, 65,
                                                     74
Plunger   62
Point   37
Poison 3, 5
Polarimeter 71, 72
Policeman 26, 32
Portable Laboratory Furniture   9
Potassium Hydroxide 42, 44
Powerstat   15
Press 85, 86
Pressure Control Valve   18
Pressure Gauge 18, 20, 21
Pressure Indicator   18
Pressure Pump 20, 22
Pressure Regulator 18, 19
Probe   114
Pulley 83, 89
Pump 20, 22, 66, 67
Punch   58
Purity Grade   43
Pycnometer 70, 72

- Q -

Quebec Colony Counter 121, 123
Quickfit 51, 55

- R -

Rack 124, 125, 126
Radiation Hazard 3, 5
Reagent 42, 44
Reagent Label 43, 44
Receptacle   15
Recorder 102, 103
Reference Electrode   87
Refractometer 71, 72
Refrigerator 3, 5, 23, 25
Regulator 18, 19
Respirator 4, 5
Respirometer 111, 112
Retort Stand 46, 47, 49
Ring Stand 46, 47, 49
Rod 50, 55
Rotary Oil Pump   20
Rotor 90, 93
Rubber Bulb 29, 32
Rubber Gloves 28, 32
Rubber Stopper   57
Rubber Tubing 60, 61
Rule 38, 39, 114
Ruler 38, 39

- S -

Safety Data   43
Safety Equipment 1, 5
Safety Goggles 4, 5
Safety Hand Grip 53, 55
Safety Matches 78, 80
Sample Holder   91
Sampler 95, 103
Scale 68, 72
Scalpel   114
Scanning Densitometer 94, 103
Scissors 33, 36, 114
Scoop 26, 32

Screwdriver 34, 36
Screw Locking Device 51
Seal 20
Seat 9
Separating Funnel 95, 103
Separation Equipment 94, 103
Separatory Funnel 95, 103
Shaker 82, 86
Shears 33, 36
Shelf 6, 11
Shower 1, 5
Side Arm 79, 105, 108
Sieve 95, 103
Signs 2, 5
Sink 11, 14
Siphon 13, 14
Size 43
Slide 116, 119
Small Animal Equipment 113
Socket 15, 17
Sodium Hydroxide 42, 44
Soldering Gun 35, 36
Soldering Iron 35
Solid Carbon Dioxide Storage Container . . 23, 24, 25
Source 43
Soxhlet Extractor 97, 103
Spatula 26, 32
Specific Gravity Balance 69, 72
Specific Gravity Bottle 70, 72
Spectro Cell 91, 93
Spectrophotometer 91, 93
Speed 81, 89
Speed Control 82, 89
Speed Indicator 89
Spout 56
Staining Dish 118, 119
Staining Jar 118, 119
Stand 46, 49
Standard Taper Ground Glass Joint 51, 55
Standard Taper Joint 52, 57
Station 118
Steam 12, 14
Step 9
Steps 9, 10
Sterile 3, 5
Sterilizer 77, 80, 121, 123
Still 108, 110
Stirrer 81, 86
STM 12
Stool 9, 10
Stopcock 12, 14, 52, 55
                                              65, 67

Stopcock Grease 65, 67
Stopper 57, 59
Stopwatch 31, 32
Storage Cabinet 6, 8
Strap 4, 18
Suction   63
Sulfuric Acid 42, 44
Sulphuric Acid 42, 44
Support 46, 49
Sweep Hand   31
Switch   15
Syringe Pipet 62, 67
Syringe Pipette 62, 67

- T -

Table 7, 8, 114, 115
Tank 18, 19
Tap 11, 14
Tape Dispenser 37, 39
Tappets   30
Telephone 2, 5
Temperature Control Knob 73, 97
Temperature Control Unit   75
Terminal   16
Test Paper 88, 93
Test Tube 27, 52, 53, 55
Test Tube Holder 53, 55
Test Tube Rack 53, 55
Thermometer 76, 80, 89, 99
Thermometer (Electronic) 76, 80
Thermos Flask 24, 25
Thermostat   74
Thimble 97, 103
Thin Layer Chromatography (TLC) 96, 103
Thumb   28
Thumbscrew 47, 48
Time   30
Timer 30, 32
Time Switch 30, 32
Tin Snips 33, 36
Tip   35
Tissue 29, 32, 117, 119
Tissue Grinder 83, 86
Tissue Processing Unit 118, 119
TLC 96, 103
Tongs 26, 32
Top 23, 24
Top-Loading Balance 69, 72
Transfer Pipet 62, 67
Transfer Pipette 62, 67

Transformer 15, 17
Trap 11
Trigger 1, 35
Triple Beam Balance 68, 72
Tripod 47, 49
Tripper 30
Trolley 9, 10, 18, 19
Truck 18
Tube 79, 80, 90, 93,
                                                 113, 115, 122,
                                                 123
Tubing 50, 54, 55, 60,
                                                 61
Tubing Clamp 60, 61
Tweezers 69

- U -

Under High Pressure 3, 5
Under Low Pressure 3, 5
Use In Hood 3, 5
Universal Indicator Paper 88, 93

- V -

VAC 13
Vacuum 13, 14, 20, 21,
                                                 22, 106
Vacuum and Air Pumps 20, 22
Vacuum and Pressure Pump 20, 22
Vacuum Gauge 20, 73, 100
Vacuum Inlet 20
Vacuum Oven 100, 103
Vacuum Pump 20, 22, 100
Valve 13, 14, 18, 20,
                                                 66
Vane Pump Movement 20
Variable Transformer 15, 17
Variac 15, 17
Vent 63
Vial 57, 59
Vice 35, 36
Viscometer 70, 72
Vise 35, 36
Voltage Control Knob 15
Volt-Ohm-Ammeter 88, 93

Volumetric Flask 66, 67
Volumetric Pipet 62, 67
Volumetric Pipette 62, 67

- W -

Wall Cabinet   6
Wall Clock 30, 32
Warning Labels 2, 5
Warning Signs 2, 5
Wash Bottle 28, 32
Washer 124, 126
Washing 124, 126
Watch Glass 53, 55
Water 1, 12, 14, 52,
                                              108
Water Bath 75, 80
Water Discharge   21
Water Pressure   21
Water Pump 21, 22
Water Still 108, 110
Wavelength Adjustment Knob   91
Wax Pencil 38, 39
Wear Your Goggles 3, 5
Weight and Sample Composition   68
Weight Control Knob   68
Weights   68
Weight Set 69, 72
Wheel   9
Wick   78
Wire Cutter Pliers 33, 36
Work Top   11
Wrench 34, 36

- Z -

Zero Reset Knob   121

中 文 索 引

— A —

石棉手套 ••••••••••••••••••••• 28、32

醋酸；乙酸 ••••••••••••••••••• 42、44

酸 ••••••••••••••••••••••••• 42、44

ＡＣＳ（美國化學學會）••••••••••• 43

可調節扳鉗 ••••••••••••••••••• 34、36

調節鈕 •••••••••••••••••••••• 34

空 氣 •••••••••••••••• 12、12、14、20、22

空氣調節 ••••••••••••••••••••• 77

空氣過濾器 ••••••••••••••••••• 20

酒 精 •••••••••••••••••••••• 42、44

酒精燈 •••••••••••••••••••••• 78、80

鋁箔盤（皿、碟）••••••••••••••• 99、103

鋁箔盤（皿、碟）••••••••••••••• 99、103

安培計 •••••••••••••••••••••• 88、93

氫氧化銨 ••••••••••••••••••••• 42、44

安培計 •••••••••••••••••••••• 16

動物籠（檻）•••••••••••••••••• 113、115

動物手術檯 ••••••••••••••••••• 114、115

圍 裙 ••••••••••••••••••••• 27、32

動脈鉗（鑷子）••••••••••••••••• 114

吸引器；吸出器（抽氣泵浦）••••••••• 21、22、101

原 子 •••••••••••••••••••••• 45

壓熱器；高壓蒸氣釜、高壓殺菌器 •••••••• 77、80

自動取樣機・・・・・・・・・・・・・・・95、103

自動傳遞吸管・・・・・・・・・・・・・・・63、67

自動傳遞吸管・・・・・・・・・・・・・・・63、67

自動歸零吸管・・・・・・・・・・・・・・・65、67

自動歸零吸管・・・・・・・・・・・・・・・65、67

避免和皮膚接觸・・・・・・・・・・・・・・3、5

安培－伏特－歐姆（ AVO ）計・・・・・・・88、93

― B ―

巴布科克（ 貝氏 ）瓶・・・・・・・・・・98、103

巴布科克（ 貝氏 ）遠心分離・・・・・・・・98、103

背；後面・・・・・・・・・・・・・・・・・9

袋　子・・・・・・・・・・・・・・・・27、32

水斗、汲出、鍋槳・・・・・・・・・・・・・56

天平、秤・・・・・・・・・・・・68、72、100、103

球窩接頭・・・・・・・・・・・・・・・51、55

原子筆・・・・・・・・・・・・・・・・37、39

氣壓計・・・・・・・・・・・・・・・・88、93

鹼、塩基、底・・・・・・・・・・・・・・44

鹼單位、基本單位・・・・・・・・・・・6、8、11

盆、皿・・・・・・・・・・・・・・・・・11

浴、浴器・・・・・・・・・・・・75、79、80、118

電池組・・・・・・・・・・・・・・・・16、17

電池充電器・・・・・・・・・・・・・・16、17

電池連接（ 接頭 ）・・・・・・・・・・・・16

燒　杯・・・・・・・・・・・・・56、59、64、67

皮　帶・・・・・・・・・・・・・・・・83、89

生物危險性 · 2、5

鑽頭；鑽孔器 · · · · · · · · · · · · · · · · · · · 35、36

毛毯；掩蓋 · 1、5

摻和器；攪拌器 · · · · · · · · · · · · · · · · · 82、86

篩；螺栓聯繫 · · · · · · · · · · · · · · · · · · · 15

彈；炸彈 · 89

彈式熱量計 · · · · · · · · · · · · · · · · · · · 88、93

結合；鍵 · 45

岩瘤頭 · 47、47、48

瓶 · · · · · · · · · · 28、32、56、59、70、72、98、103、120、123

箱；盒 · · · · · · · · · · · · · · · · · 24、25、78、121、123

靫；制動器 · · · · · · · · · · · · · · · · · · 98

鬃 毛 · 29

刷 子 · 29、32

瓷過濾漏斗 · · · · · · · · · · · · · · · · · · 105、107

斗；桶；吊桶 · · · · · · · · · · · · · · · · · · 56、59

緩衝液 · 87、93

球管；球；燈泡 · · · · · · · · · · · · · · · 29、32、63、67

滴定管；量管 · · · · · · · · · · · · · · · · · 64、67

滴定管夾 · 49、49

滴定管架 · 46

滴定管 · · · · · · · · · · · · · · · · · · 49、64、65、67

滴定管夾 · 47、49

滴定管架 · 46

燈頭、爐；燃燒器 · · · · · · · · · · · · · · · 77、80、108

— C —

櫃；櫥 · 6、7、8、96

籠；檻 · 113、115

氫氧化鈣 · 42、44

計算機 · 38、39

校準環 · 66

熱量計；熱量器 · · · · · · · · · · · · · · · · · 88、93

氫氧化鈣 · 42

蓋；帽 · · · · · · · · · · · · · 18、29、57、59、78、122

毛細管 · 50、55

二氧化碳 · · · · · · · · · · · · · · · · · · · 23、24、25

載 體 · 90、93

三輪車；用車運送 · · · · · · · · · · · · · · · · 9、10

盒；箱 · 7、8

目錄號碼 · 43

苛性的；腐蝕的；焦散的 · · · · · · · · · · · · 3、5

謹慎；小心 · 2、5

小室；電池；細胞 · · · · · · · · · · · · · · · · 91、93

遠心分離機 · · · · · · · · · · · · · · · 88、93、98、103

離心機轉頭 · 90、93

離心機轉動子（轉座）· · · · · · · · · · · · · · 90、93

離心管 · 90、93

檢定的；證明的 · · · · · · · · · · · · · · · · · · 43

鏈 · 114

椅 子 · 9

充電器；裝料機 · · · · · · · · · · · · · · · · · 16、17

圖；圖表；表 · · · · · · · · · · · · · · · · · · · 40、41

酒精；乙醇 · 42

醋酸；乙酸 · 42

抑止螺旋 · · · · · · · · · · · · · · · · · · · 48

色層分析法 · · · · · · · · · · · · · · · 96、103

色層分析櫥 · · · · · · · · · · · · · · · · · 96

色層分析柱 · · · · · · · · · · · · · · · 96、103

夾；夾板、壓板 · · · · · · · 18、19、47、48、49、51、55、60、61、83

持夾把手、持夾器 · · · · · · · · · · · · · · 47、49

附把手夾 · · · · · · · · · · · · · · · · 48、49

清潔劑 · · · · · · · · · · · · · · · · · 29、32

夾；紙夾 · · · · · · · · · · · · · · · · 37、65

鐘；計時器 · · · · · · · · · · · · · · · 30、32

冷　藏 · · · · · · · · · · · · · · · · · 23、25

冷　水 · · · · · · · · · · · · · · 12、14、52、108

收集瓶 · · · · · · · · · · · · · · · · · · 108

菌落計數器 · · · · · · · · · · · · · · · 121、123

柱、塔 · · · · · · · · · · · · · · · · · 96、103

同心環 · · · · · · · · · · · · · · · · · · 75

冷凝器；電容器；聚光器 · · · · · · 52、55、97、100、101、108

錐形瓶 · · · · · · · · · · · · · · · · · 56、59

容器和密閉器 · · · · · · · · · · · · · · · 56、59

軟木塞；軟木；塞 · · · · · · · · · · · · · 58、59

穿孔器 · · · · · · · · · · · · · · · · · 58、59

腐蝕劑；腐蝕的 · · · · · · · · · · · · · · · 3、5

計數器；逆；反 · · · · · · · · · · · · · · 121、123

計算室 · 117、119

粗調節螺 · 116

曲 柄 · 89、118

新月形扳鉗 · · · · · · · · · · · · · · · · · · · 34、36

燈帽；窯頂 · 31

坩堝；熔鍋 · · · · · · · · · · · · · · · 75、80、105、107

坩堝座 · 106、107

培養基 · 120、123

培養管 · 122、123

杯 櫥 · 3、5、6、8

切刀；切斷機 · · · · · · · · · · · · · · · · · 37、54、55

C W（冷水） · · · · · · · · · · · · · · · · · · 12

圓筒、量筒、氣缸 · · · · · · · · · · · · · 18、19、63、67

圓筒聯合（連結） · · · · · · · · · · · · · · · 18

氣缸氣體；鋼瓶氣 · · · · · · · · · · · · · · · 18

— D —

危 險 · 3、5

脫礦物質器（劑） · · · · · · · · · · · · · · 109、110

深淺計；光密度計 · · · · · · · · · · · · · · 94、103

除濕乾燥器 · · · · · · · · · · · · · · · · · · 97、103

桌子、實驗檯 · · · · · · · · · · · · · · · · · · 7、8

真空瓶；杜瓦瓶 · · · · · · · · · · · · · · · 24、25

刻度盤；標度 · · · · · · · · · · · · · · 2、68、76、80

刻度溫度計 · · · · · · · · · · · · · · · · · · · 76、80

滲析膜；滲析器・・・・・・・・・・・・・・・・・・・・94、103

稀釋瓶・・・・・・・・・・・・・・・・・・・・・・・120、123

碟、皿、盤・・・・27、32、99、103、113、115、118、119、120、121、123

使用後丟棄手套・・・・・・・・・・・・・・・・・・・28、32

解剖用具（組、套）・・・・・・・・・・・・・・・・114、115

解剖用具組・・・・・・・・・・・・・・・・・・・114、115

蒸餾作用・・・・・・・・・・・・・・・・・・・・108、110

蒸餾瓶・・・・・・・・・・・・・・・・・・・・・108、110

蒸餾水・・・・・・・・・・・・・・・・・・・・・12、14

蒸餾瓶；分餾燒瓶・・・・・・・・・・・・108、108、110

分配板・・・・・・・・・・・・・・・・・・・・・・15

分配槽・・・・・・・・・・・・・・・・・・・・15、17

分規；雙脚規・・・・・・・・・・・・・・・・・98、103

門・・・・・・・・・・・・・・・・・・・・・・6、23

排水架；滴乾架・・・・・・・・・・・・・・・12、5、126

排水管；洩水管・・・・・・・・・・・・・・・・・124

抽屜、衣櫃・・・・・・・・・・・・・・・・・・・6

鑽孔器、螺旋錐・・・・・・・・・・・・・・・35、36

吸水管・・・・・・・・・・・・・・・・・・・113、115

滴　瓶・・・・・・・・・・・・・・・・・・・29、32

乾氷浴槽・・・・・・・・・・・・・・・・・・・100

乾氷盒・・・・・・・・・・・・・・・・・・・24、25

乾氷貯藏器（容器）・・・・・・・・・・・・・・23、25

乾燥機、烘乾機・・・・・・・・・・・・・・・99、103

ＤＷ（蒸餾水）・・・・・・・・・・・・・・・・・12

— E —

電氣的・・・・・・・・・・・・・・・・・・・・・・・・・・15、17

電的出路，電路・・・・・・・・・・・・・・・・・・・・15、17

電插頭・・・・・・・・・・・・・・・・・・・・・・・・・・15、17

電　極・・・・・・・・・・・・・・・・・・・・・・・・87、93

電　泳・・・・・・・・・・・・・・・・・・・・・・・94、103

乳濁液均質機（乳化機）・・・・・・・・・・・・・・84、86

Erlenmeyer 瓶（細頸三角瓶）・・・・・・・・・・56、59

浸蝕區域標誌・・・・・・・・・・・・・・・・・・・・・・120

蒸發皿・・・・・・・・・・・・・・・・・・・・・・・・27、32

蒸發器・・・・・・・・・・・・・・・・・・・・・・109、110

爆炸性的・・・・・・・・・・・・・・・・・・・・・・・・2、5

萃取殼筒・・・・・・・・・・・・・・・・・・・・・97、103

萃取管・・・・・・・・・・・・・・・・・・・・・・・・・・97

萃取器・・・・・・・・・・・・・・・・・・・・97、97、103

目　鏡・・・・・・・・・・・・・・・・・・・・・・・・・116

洗眼水・・・・・・・・・・・・・・・・・・・・・・・・・2、5

— F —

水龍頭管・・・・・・・・・・・・・・・・・・・・・・・11、14

送料皿（碟）・・・・・・・・・・・・・・・・・・・113、115

簽字筆（細筆尖）・・・・・・・・・・・・・・・・・38、39

銼　刀・・・・・・・・・・・・・・・・・・・・・・・・35、36

銼屑櫥・・・・・・・・・・・・・・・・・・・・・・・・・7、8

裝油管線・・・・・・・・・・・・・・・・・・・・・・・・120

吸濾瓶・・・・・・・・・・・・・・・・・・・・・・105、107

過　濾・・・・・・・・・・・・・・・・・・・・104、105、107

吸濾瓶 · 105、107

濾　紙 · 104、107

細調節螺 · 116

細鼻鉗 · 33

手　指 · 28、28、28

防火毯 · 1、5

滅火器 · 1、5

急救箱 · 1、5

裝置品；裝置器 · · · · · · · · · · · · · · · · 11

可燃的 · 3、5

急驟蒸發器 · · · · · · · · · · · · · · · · · · · 109、110

燒瓶；三角瓶24、25、56、57、59、66、67、97、101、103、105、107、108、110

浮標；浮筒；浮 · · · · · · · · · · · · · · · · · 92

落地櫥 · 7·8

平底燒瓶 · 57、59

流量的控制 · · · · · · · · · · · · · · · · · · · 13

流速計、流量計 · · · · · · · · · · · · · · · · 92、93

氟量計；螢光計 · · · · · · · · · · · · · · · · 92、93

氟量計；螢光計 · · · · · · · · · · · · · · · · 92、93

箔；薄片、鋁箔片 · · · · · · · · · · · · 27、32、99、103

箔片帽（蓋）· · · · · · · · · · · · · · · · · 27

箔片皿（碟）· · · · · · · · · · · · · · · · · 27

鑷　子 · 114

框　架 · 46、49

冷凍乾燥 · 100、103

冷凍機 · 23、25

烟橱（烟室）· 6、8

烟橱；通風室、抽油煙機· · · · · · · · · · · · · · · 6、8

漏　斗· · · · · · · · · · · · · · · · 95、103、104、105、107

爐· 75、80

傢　具· 6、9

保險絲；引信、導火線· · · · · · · · · · · · · · · 15、15

— G —

氣　體· · · · · · · · · · · · · 13、14、18、19、78、80

空氣調節· 77

氣體鋼瓶· 18

氣體鋼瓶夾（轉栓）· · · · · · · · · · · · · · 18、19

氣體打火器· · · · · · · · · · · · · · · · · · · 78、80

氣體測定器· · · · · · · · · · · · · · · · · 111、112

表；規、計· · · · · · · · · · · · · · 18、20、21、85

膠體電泳柱（塔）· · · · · · · · · · · · · · 94、103

玻璃；眼鏡· · · · · · · · · · · · · · · · · · 50、55

玻璃電極· 87

玻璃棒· 50、55

玻璃塞· 52、55

玻璃塞· 57、59

玻璃管· 50、55

玻璃管切斷器· · · · · · · · · · · · · · · · 54、55

玻璃器皿洗滌機· · · · · · · · · · · · · · 124、126

玻璃棉；玻璃絨· · · · · · · · · · · · · · 104、107

手 套 · · · · · · · · · · · · · · · · · · 28、32

眼罩；護目眼鏡 · · · · · · · · · · · · · · 3、4、5

金鈞鈎分離器（Gold Fisch ） · · · · · · · · · · 97、103

量筒；量杯、刻度 · · · · · · · · · · · · · 64、67

刻度滴定管 · · · · · · · · · · · · · · · · 64、67

刻度滴定管 · · · · · · · · · · · · · · · · 64、67

刻度量筒 · · · · · · · · · · · · · · · · · 63、67

刻度吸管 · · · · · · · · · · · · · · · · · 62、67

刻度吸管 · · · · · · · · · · · · · · · · · 62、67

潤滑脂；膏脂 · · · · · · · · · · · · · · · 65、67

碾磨；碾碎 · · · · · · · · · · · · · 83、83、86

— H —

錘 子 · · · · · · · · · · · · · · · · · · 34、36

手 · · · · · · · · · · · · · · · · · · · 30、31

手動碾磨器 · · · · · · · · · · · · · · · · 83、86

手動切碎（切片）器 · · · · · · · · · · · · 83、86

實驗室使用設備 · · · · · · · · · · · · · · 26、32

手動的；可使用的、柄；把手 · · · · 7、21、26、29、34、56、83、85

手動泵（唧筒） · · · · · · · · · · · · · 21、22

手動泵（地板式） · · · · · · · · · · · · · 21、22

手提式反射計 · · · · · · · · · · · · · · · 71、72

氯化氫、塩酸 · · · · · · · · · · · · · · · · 42

加熱器 · · · · · · · · · · · · · · · · · 101、101

加 熱 · · · · · · · · · · · · · · · · · 73、80

加熱罩 · · · · · · · · · · · · · · · · · 74、80

加熱帶 · · · · · · · · · · · · · · · · · 74、80

六角形扳鉗・・・・・・・・・・・・・・・・・・・・・・34、36

硝　酸・・・・・・・・・・・・・・・・・・・・・・・・　42

座；支架：柄；把持・・・・・・・・・・・・・・・・・・　122

洞；孔；穴・・・・・・・・・・・・・・・・・・・・・　57

均質機；均化器・・・・・・・・・・・・・・・・・・84、86

通風櫥；蓋、抽油煙機・・・・・・・・・・・・・・3、5、6、8

鈎・・・・・・・・・・・・・・・・・・・・・・・・・　114

軟管連接・・・・・・・・・・・・・・・・・・・・・・　18

熱的・・・・・・・・・・・・・・・・・・・・・・・・3、5

熱板；電爐・・・・・・・・・・・・・・・・・・・・73、80

熱板高度調節螺・・・・・・・・・・・・・・・・・・・　97

熱　水・・・・・・・・・・・・・・・・・・・・・・12、14

硫　酸・・・・・・・・・・・・・・・・・・・・・・・　42

ＨＷ（熱水）・・・・・・・・・・・・・・・・・・・　12

水壓機・・・・・・・・・・・・・・・・・・・・・・85、86

塩　酸・・・・・・・・・・・・・・・・・・・・・・42、44

比重計・・・・・・・・・・・・・・・・・・・・・・70、72

氫氧化物・・・・・・・・・・・・・・・・・・・・・42、44

　　－ I －

浸液加熱器・・・・・・・・・・・・・・・・・・・・74、80

衝擊器；震動器・・・・・・・・・・・・・・・・・106、107

孵卵器・・・・・・・・・・・・・・・・・・・・・・73、80

入口管；入口・・・・・・・・・・・・・・・・・・・　20

入口管・・・・・・・・・・・・・・・・・・・・64、65、124

播種環（圈）、接種環、白金耳・・・・・・・・・・・・・・・122、123
接種針・・・・・・・・・・・・・・・・・・・・・・・・122、123
消毒（殺菌）器・・・・・・・・・・・・・・・・・・・・・77、80
進入轉栓（活塞）・・・・・・・・・・・・・・・・・・・・・ 13
間隔時鐘（計時器）・・・・・・・・・・・・・・・・・・・30、32

— J —

缸；瓶・・・・・・・・・・・・・・・・・・・・・・・118、119
接頭；接合；關節・・・・・・・・・・・・・・・・・・・51、55

— K —

保存在冷却器內（冷却冷藏）・・・・・・・・・・・・・・・ 3、5
凱氏分解裝置・・・・・・・・・・・・・・・・・・・・101、103
凱氏蒸餾裝置・・・・・・・・・・・・・・・・・・・・101、103
凱氏燒瓶・・・・・・・・・・・・・・・・・・・・・・101、103
刀；小刀・・・・・・・・・・・・・・・・・・・・・・・ 118
氫氧化鉀・・・・・・・・・・・・・・・・・・・・・・・ 42

— L —

標　識・・・・・・・・・・・・・・2、5、25、37、39、43、44
標識號誌・・・・・・・・・・・・・・・・・・・・・・・37、39
實驗室推車（吊車）・・・・・・・・・・・・・・・・・・9、10

實驗室椅子 · 9
實驗室用具（家具） · · · · · · · · · · · · · · 6
實驗室使用工具 · · · · · · · · · · · · · · · · 33
實驗室凳子 · · · · · · · · · · · · · · · · · · · 9、10
實驗室手推車 · · · · · · · · · · · · · · · · · 9、10
梯　子 · 9、10
燈 · 78、80
大型搖盪機 · · · · · · · · · · · · · · · · · · · 82
支架；腿 · 7
鏡頭擦拭紙 · · · · · · · · · · · · · · · · · 117、119
槓桿；桿 · 2、30
蓋子控制把 · · · · · · · · · · · · · · · · · · 77
發光器；點火器 · · · · · · · · · · · · · · · 78、80
光　源 · 116
唇；口緣 · 56
液體體積測定設備 · · · · · · · · · · · · · · 62
石蕊試紙 · · · · · · · · · · · · · · · · · · · 88、93
開　鎖 · 7
迴線；線、鉤 · · · · · · · · · · · · · · · · · 122、123
整批分析 · 43
批　號 · 43

— M —

磁鐵攪動器 · · · · · · · · · · · · · · · · · · 81、86

放大鏡 · · · · · · · · · · · · · · · · 116、119、121

罩；紗罩 · · · · · · · · · · · · · · · · · · 74、80

火　柴 · · · · · · · · · · · · · · · · · · · 78、80

Mcleod 真空計 · · · · · · · · · · · · · · · · 100

測量用具 · · · · · · · · · · · · · · · · · 87、93

量液吸管 · · · · · · · · · · · · · · · · · 62、67

量液吸管 · · · · · · · · · · · · · · · · · 62、67

機械分段控制 · · · · · · · · · · · · · · · · 116

培養基 · · · · · · · · · · · · · · · · · 120、123

熔點浴 · · · · · · · · · · · · · · · · · · 79、80

熔點管 · · · · · · · · · · · · · · · · · · 79、80

金屬銼刀 · · · · · · · · · · · · · · · · · 35、36

米制的；度量的、公制的 · · · · · · · · · · · · 43

微生物學 · · · · · · · · · · · · · · · · 120、123

測微計 · · · · · · · · · · · · · · · · · 117、119

顯微鏡 · · · · · · · · · · · · · · · · · 116、119

顯微鏡檢查 · · · · · · · · · · · · · · · 116、119

切片機 · · · · · · · · · · · · · · · · · 118、119

磨；碾機；壓碎 · · · · · · · · · · · · · · 83、86

絞碎器 · · · · · · · · · · · · · · · · 83、83、86

混合水龍頭 · · · · · · · · · · · · · · 11、11、14

混合和加壓設備 · · · · · · · · · · · · · · 81、86

混合水龍頭 · · · · · · · · · · · · · · 11、11、14

模式；模型 · · · · · · · · · · · · · · · · · 45

水分測定計 · 100、103

分子模型 · 45

碾鉢和杵 · 81、86

電動機；馬達 · · · · · · · · · · · · · · · · · 20、83、89

蒙烰爐；隔焰爐；迴熱爐 · · · · · · · · · · · · · · 75、80

— N —

氫氧化鈉 · 42

針鼻鉗 · 33

針狀塞（栓）、針孔閥 · · · · · · · · · · · · · · 13、14、18、20

氫氧化氨 · 42

螺紋接管 · 21

NIT（氮氣） · · · · · · · · · · · · · · · · · · · 13

硝　酸 · 42、44

氮　氣 · 13、14

筆記簿 · 26、32

螺母、螺套 · 52

— O —

物鏡；目的；目標 · · · · · · · · · · · · · · · · · 116

辦公室使用工具 · · · · · · · · · · · · · · · · · 37、39

歐　姆 · 88、93

油　面 · 20

集油槽；油收集器 · · · · · · · · · · · · · · · · · 20

轉換開關 · · · · · · · · · · · · · · · · · · · 15、15、121

開口端扳鉗 · · · · · · · · · · · · · · · · · 34、36

操作中；操作的；工作的；外科手術的 · · · · · · · · 114、115

O－環聯接 · · · · · · · · · · · · · · · · 51、55

OSHA · · · · · · · · · · · · · · · · · · · 43

出口管；出口 · · · · · · · · · · · · · · · 15、15、17、20

出口箱；出口室 · · · · · · · · · · · · · · · 15

出口窄道（配線箱）· · · · · · · · · · · · · · 15、17

失靈、壞掉 · · · · · · · · · · · · · · · · 3、5

爐；烘乾箱 · · · · · · · · · · · · · 73、80、99、100、103

溢　出 · · · · · · · · · · · · · · · · · · 11、13

溢流杯 · · · · · · · · · · · · · · · · · 64、65

氧　氣 · · · · · · · · · · · · · · · · · 13

氧氣（O₂）· · · · · · · · · · · · · · · · 13、14

－ P －

提桶；桶 · · · · · · · · · · · · · · · · 56、59

盤；鍋 · · · · · · · · · · · · · · · · 68、68、69

筆；檻 · · · · · · · · · · · · · · · · · 37、39

週期表 · · · · · · · · · · · · · · · · · 40、41

週期表 · · · · · · · · · · · · · · · · · 40、41

杵 · · · · · · · · · · · · · · · · · · · 81、86

培養皿 · · · · · · · · · · · · · · · · · 120、123

培養皿殺菌盒（筐）· · · · · · · · · · · · · 121、123

ＰＨ ；酸鹼度・・・・・・・・・・・・・・・・・・・87、88、93

酸鹼緩衝液・・・・・・・・・・・・・・・・・・・・87、93

ＰＨ 測定計・・・・・・・・・・・・・・・・・・・87、93

電　話・・・・・・・・・・・・・・・・・・・・・・2

引火燈・・・・・・・・・・・・・・・・・・・・・15

挾夾；揑夾・・・・・・・・・・・・・・・・・・51、55

吸　管・・・・・・・・・・・・・・・29、62、63、65、67、126

附球充填吸管・・・・・・・・・・・・・・・・・63、67

吸管；吸移管；吸量管・・・・・・・29、32、62、63、65、67、126

附充填器吸量管・・・・・・・・・・・・・・・・63、67

附球充填吸管・・・・・・・・・・・・・・・・・63、67

吸管洗滌器・・・・・・・・・・・・・・・・・124、126

吸管洗滌器・・・・・・・・・・・・・・・・・124、126

塑膠塞子・・・・・・・・・・・・・・・・・・・57

塑膠試管・・・・・・・・・・・・・・・・・・60、61

扁嘴鉗；手鉗・・・・・・・・・・・・・・・・33、36

管塞；塞子；插頭・・・・・・・・・・15、16、52、65、74

柱　塞・・・・・・・・・・・・・・・・・・・・62

點；尖端；要點・・・・・・・・・・・・・・・・37

毒；毒物；毒藥、中毒・・・・・・・・・・・・・3、5

旋光計；偏振計・・・・・・・・・・・・・・・71、72

澱帚・・・・・・・・・・・・・・・・・・・・26、32

手提式實驗室用具・・・・・・・・・・・・・・・9

氫氧化鉀・・・・・・・・・・・・・・・・・・42、44

發電廠；動力廠・・・・・・・・・・・・・・・・15

壓機；壓力機；壓榨機・・・・・・・・・・・・85、86

壓力控制閥・・・・・・・・・・・・・・・・・・18

壓力計・・・・・・・・・・・・・・・・・18、20、21

壓力指示器・・・・・・・・・・・・・・・・・・18

壓氣泵・・・・・・・・・・・・・・・・・・・20、22

調壓器・・・・・・・・・・・・・・・・・・・18、19

試探具；試探電極；探針 · · · · · · · · · · 114

滑輪；滑車；皮帶輪 · · · · · · · · · 83、89

泵浦；唧筒；抽機 · · · · · · · · 20、22、66、67

衝床；沖壓機；穿孔器 · · · · · · · 58

純度分級 · · · · · · · · · · 43

比重瓶 · · · · · · · · · 70、72

— Q —

Quebec 菌落計數器 · · · · · · · · 121、123

快速裝配 · · · · · · · · 51、55

— R —

架；齒條 · · · · · · · · · · 124、125、126

幅射危險性 · · · · · · · · 3、5

試劑；試藥 · · · · · · · · 42、44

試藥標籤 · · · · · · · · 43、44

收容器；插座 · · · · · · · 15

記錄器 · · · · · · · · · 102、103

參考電極 · · · · · · · 87

折射計；折光儀 · · · · · · · 71、72

冷凍器（機）；冷藏箱 · · · · · · 3、5、23、25

調節劑；調節器 · · · · · · · 18、19

面罩；呼吸器；濾毒罐 · · · · · · 4、5

呼吸計 · · · · · · · · 111、112

瓶架；瓶餾座 · · · · · · 46、47、49

環架；鐵環架 · · · · · · · 46、47、49

棒條；圓條 · · · · · · · 50、55

旋轉石油泵（幫浦） · · · · · 20

轉子；轉動片、旋度 · · · · · · 90、93

橡皮吸球 ·29、32

橡皮手套 ·28、32

橡皮塞子（栓）· · · · · · · · · · · · · · · · · · · 57

橡皮（膠）管 · · · · · · · · · · · · · · · · · · ·60、61

尺、定則、法則 · · · · · · · · · · · · · · · ·38、39、114

尺；米達尺 ·38、39

— S —

安全數據 · 43

安全裝備 · 1、5

安全護目鏡 · 4、5

安全把手 ·53、55

安全火柴 ·78、80

樣品固定器（試管架）· · · · · · · · · · · · 91

採樣器 · 95、103

秤 ·68、72

外科用小刀 · · · · · · · · · · · · · · · · · · · 114

微調光密度計 · · · · · · · · · · · · · · · · · 94、103

剪　刀 ·33、36、114

勺；杓 · 26、32

螺絲起子 · 34、36

螺絲之深陷螺紋 · · · · · · · · · · · · · · · · · 51

密封 · 20

座；坐位 · 9

分液漏斗 · 95、103

分離設備 · 94、103

分液漏斗 · 95、103

搖動器、振盪器 · · · · · · · · · · · · · · · 82、86

切斷裝置 · · · · · · · · · · · · · · · · · · 33、36

架　子 · 6、11

淋浴器 · 1、5

側　臂 · · · · · · · · · · · · · · · · 79、105、108

篩　子 · 95、103

符號；標示；簽字 · · · · · · · · · · · · · · 2、5

水槽；污水槽 · · · · · · · · · · · · · 11、11、14

虹吸管；虹吸 · · · · · · · · · · · · · 13、14

大小；尺度 · · · · · · · · · · · · · · · · 43

玻　片 · · · · · · · · · · · · · · · · · · · 116、119

小動物用器具 · · · · · · · · · · · · · · 113

插　座 · · · · · · · · · · · · · · · · · 15、17

氫氧化鈉 · · · · · · · · · · · · · · · · 42、44

焊接槍；電焊槍 · · · · · · · · · · · · · 35、36

固態二氧化碳貯存器 · · · · · · · 23、24、25

來源；製造商 · · · · · · · · · · · · · 43

Soxhlet（索格利特）萃取器 · · · · · 97、103

刮勺；刮鏟 · · · · · · · · · · · · · · 26、32

比重天平 · · · · · · · · · · · · · · · · 69、72

比重瓶 · · · · · · · · · · · · · · · · · 70、72

光譜室 · · · · · · · · · · · · · · · · · 91、93

分光光度計 · · · · · · · · · · · · · · · 91、93

速率；速度 · · · · · · · · · · · · · · · · · · · 81、89

速度控制（調節）· · · · · · · · · · · · · · · 82、89

速率計；速度計；速度指示計 · · · · · · · · · 89

嘴；唇 · 56

染色皿 · · · · · · · · · · · · · · · · · · · 118、119

染色缸 · · · · · · · · · · · · · · · · · · · 118、119

架；臺 · · · · · · · · · · · · · · · · · · · 46、49

標準錐削玻璃聯結 · · · · · · · · · · · · · 51、55

標準錐削聯結 · · · · · · · · · · · · · · · 52、57

浸漬槽 · 118

蒸氣；水蒸氣 · · · · · · · · · · · · · · · 12、14

梯　階 · 9

梯　階 · · · · · · · · · · · · · · · · · · · 9、10

無菌的 · · · · · · · · · · · · · · · · · · · 3、5

殺菌器；消毒器 · · · · · · · · · · · 77、80、121、123

蒸餾釜（鍋）；蒸餾 · · · · · · · · · · · · ·108、110

攪拌器；攪動器 · · · · · · · · · · · · · · 81、86

STM（蒸氣）· · · · · · · · · · · · · · · 12

活栓脂膏、潤滑膏 · · · · · · · · · · · · · 65、67

凳子、椅子 · · · · · · · · · · · · · · · · · 9、10

活栓；轉閂；活塞 · · · · · · · · · 12、14、52、55、65、67

塞　子 · · · · · · · · · · · · · · · · · · · 57、59

碼錶；計時錶 · · · · · · · · · · · · · · · 31、32

儲存櫥 · · · · · · · · · · · · · · · · · · · 6、8

帶　子 · · · · · · · · · · · · · · · · · · · 4、18

空吸；吸取；吸力 · · · · · · · · · · · · · · · · · 63

硫　酸 · · · · · · · · · · · · · · · · · · 42、44

硫　酸 · · · · · · · · · · · · · · · · · · 42、44

架；支座 · · · · · · · · · · · · · · · · · 46、49

長針（可歸零）· · · · · · · · · · · · · · 31

電閘；開關 · · · · · · · · · · · · · · · · · 15

注射管 · · · · · · · · · · · · · · · · · · 62、67

注射管 · · · · · · · · · · · · · · · · · · 62、67

— T —

表；桌子 · · · · · · · · · · · · · · · 7、8、114、115

桶；罐；槽 · · · · · · · · · · · · · · · · · 18、19

水龍頭 · · · · · · · · · · · · · · · · · · 11、14

膠帶固定座（捲尺）· · · · · · · · · · 37、39

放液（水）管 · · · · · · · · · · · · · · 30

電話；電話機 · · · · · · · · · · · · · · · 2、5

溫度控制開關 · · · · · · · · · · · · · · · 73、97

溫度控制裝置 · · · · · · · · · · · · · · · 75

端；末端；終點 · · · · · · · · · · · · · · 16、16

試　紙 · · · · · · · · · · · · · · · · · · 88、93

試　管 · · · · · · · · · · · · · · 27、52、53、55

試管夾 · · · · · · · · · · · · · · · · · · 53、55

試管架 · · · · · · · · · · · · · · · · · · 53、55

溫度計 · · · · · · · · · · · · · · · 76、80、89、89

電子溫度計 · · · · · · · · · · · · · · · · 76、80

保溫瓶 · · · · · · · · · · · · · · · · · · 24、25

恒溫器 · · · · · · · · · · · · · · · · · · 74

殼筒；套管 · · · · · · · · · · · · · · · · 97、103

薄層色層分析（TLC）· · · · · · · · · 96、103

拇　指 · · · · · · · · · · · · · · · · 28、28、28

拇指螺絲（螺旋）· · · · · · · · · · · · 47、47、48

時　　間 ・・・・・・・・・・・・・・・・・・　30

自動定時儀・・・・・・・・・・・・・・・・・30、32

定時開關・・・・・・・・・・・・・・・・・・30、32

錫剪；錫鋏子・・・・・・・・・・・・・・・・33、36

管尖；尖・・・・・・・・・・・・・・・・・　35

擦拭紙・・・・・・・・・・・・・・・29、32、117、119

組織碾磨器・・・・・・・・・・・・・・・83、86

組織處理器位・・・・・・・・・・・・・118、119

TLC（薄層液體色析法）・・・・・・・・96、103

鉗・・・・・・・・・・・・・・・・・・・26、32

頂；上面；蓋頂・・・・・・・・・・・・・・23、24

上皿天平・・・・・・・・・・・・・・・・69、72

移量吸管・・・・・・・・・・・・・・・・62、67

移量吸管・・・・・・・・・・・・・・・・62、67

變壓器・・・・・・・・・・・・・・・・・15、17

阱；汽水閥・・・・・・・・・・・・・・・・　11

引發物；扳機・・・・・・・・・・・・・・・1、35

三標天平；三標平・・・・・・・・・・・・・68、72

三脚架・・・・・・・・・・・・・・・・・47、49

傾料器・・・・・・・・・・・・・・・・・・　30

空中吊運車・・・・・・・・・・・・9、10、18、19

貨車；卡車・・・・・・・・・・・・・・・・　18

管；筒・・・・・・79、80、90、93、113、115、122、123

管；製管・・・・・・・・・・・50、54、55、60、61

試管夾・・・・・・・・・・・・・・・・・60、61

鑷　子・・・・・・・・・・・・・・・・・・　69

在高壓情況下・・・・・・・・・・・・・・・・・・・3、5

在低壓情況下・・・・・・・・・・・・・・・・・・・3、5

在通風室內使用・・・・・・・・・・・・・・・・・・3、5

萬用（通用）試紙・・・・・・・・・・・・・・・・88、93

— V —

VAC（真空）・・・・・・・・・・・・・・・・・・・13

真　空・・・・・・・・・・・13、14、20、21、22、106

真空和空氣泵（帮浦）・・・・・・・・・・・・・20、22

真空和壓力泵（帮浦）・・・・・・・・・・・・・20、22

真空計・・・・・・・・・・・・・20、73、100、100

真空入口、抽真空口・・・・・・・・・・・・・・・・20

真空烘乾箱（器）・・・・・・・・・・・・・・100、103

真空泵（帮浦）・・・・・・・・・・・・20、22、100

活栓、閥；活塞・・・・・・・・13、14、18、20、66

移動式葉輪泵・・・・・・・・・・・・・・・・・・・20

可變變壓器・・・・・・・・・・・・・・・・・15、17

可變變壓器・・・・・・・・・・・・・・・・・15、17

孔；通氣口・・・・・・・・・・・・・・・・・63、63

管；瓶・・・・・・・・・・・・・・・・・・・57、59

軋　鉗・・・・・・・・・・・・・・・・・・・35、36

黏度計・・・・・・・・・・・・・・・・・・・70、72

老虎鉗；虎頭鉗・・・・・・・・・・・・・・・35、36

電壓控制鈕・・・・・・・・・・・・・・・・・・・15

伏特－歐姆－安培計・・・・・・・・・・・・・88、93

量　瓶・・・・・・・・・・・・・・・・・・・66、67

容量吸管・・・・・・・・・・・・・・・・・・・62、67

容量吸管・・・・・・・・・・・・・・・・・・・62、67

壁　櫥・・・・・・・・・・・・・・・・・・・・6

壁　鐘・・・・・・・・・・・・・・・・30、32

警告標誌・・・・・・・・・・・・・・・2、5

警告信號・・・・・・・・・・・・・・・2、5

洗瓶；沖洗瓶・・・・・・・・・・・・28、32

洗滌器；洗滌機・・・・・・・・・・124、126

洗　滌・・・・・・・・・・・・・・124、126

錶玻璃・・・・・・・・・・・・・・・53、55

水・・・・・・・・・・・・1、12、14、52、108

水　浴・・・・・・・・・・・・・・・75、80

排水；排水架・・・・・・・・・・・・・21

水　壓・・・・・・・・・・・・・・・・21

水　泵・・・・・・・・・・・・・・・21、22

蒸餾水製造器・・・・・・・・・・・108、110

波長調節鈕・・・・・・・・・・・・・・91

蠟　筆・・・・・・・・・・・・・・・38、39

帶上你的護目鏡・・・・・・・・・・・3、5

重量和樣品組成・・・・・・・・・・・・68

重量調節（控制）鈕・・・・・・・・・・68

法　碼・・・・・・・・・・・・・・・・68

法碼組・・・・・・・・・・・・・・・69、72

輪子・・・・・・・・・・・・・・・・・9

渦轉式混合器・・・・・・・・・・・・・82

蕊；燈心・・・・・・・・・・・・・・・78

剪鐵絲鉗・・・・・・・・・・・・・・33、36

工作臺・・・・・・・・・・・・・・・・11

扳　鉗・・・・・・・・・・・・・・・34、36

- A -

Acide acétique 42
Acide hydrochlorique 42
Acide nitrique 42
Acide sulfurique 42
Affiches 2
Agitateur 26, 81
Agitateur magnétique 81
Aiguille d'inoculation 122
Air 12
Alambic 108
Alambic à eau 108
Alcool 42
Alumettes 78
Aluminium, assiette en papier 99
Aluminium, papier 27
Ampoule en caoutchouc 29
Appareil de papier collant 37
Appareil respiratoire 111
Aspirateur (pompe à eau) 21
Assiette 113
Assiette d'évaporation 27
Autoclave ou stérilisateur 77

- B -

Baguette de verre 50
Bain à eau 75
Bain à point de fusion 79
Balance 68
Balance à densité spécifique 69
Balance à trois fléaux 68
Balance électronique 69
Ballon filtre 105
Baromètre 88
Basse pression, sous 3
Batterie, chargeable et non-rechargeable 16
Bécher 56, 64
Boîte de stérilisation du plat de Petri 121
Bouchon 58
Bouchon d'arrêt 12, 65
Bouchon d'arrêt en caoutchouc ou plastique 57
Bouchon d'arrêt en verre 57
Bouteille 56
Bouteille à densité spécifique 70
Bouteille à gaz 18
Bouteille de Babcock 98
Bouteille de dilution 120
Briquet à gaz 78
Brosse 29
Broyeur à main 83

Broyeur de tissus   83
Brûleur   77
Bureau   7
Burette graduée   64

- C -

Cabinet au sol   7
Cage   113
Calculatrice   38
Calorimètre   89
Canne de verre   50
Carnet   26
Cartouche de Soxhlet   97
Cartouche d'extraction   97
Caustique   3
Centrifugeuse   89
Centrifugeuse de Babcock   98
Centrifugeuse, tête de   90
Centrifugeuse, tube de   90
Charnière   51
Charriot de laboratoire   9
Chaud   3
Chauffage d'immersion   74
Chromatographie à couche mince   96
Chromatographie, cabinet de   96
Chromatographie, colonne de   96
Chronomètre   31
Clé anglaise   34
Clé anglaise à bout ouvert   34
Clés anglaises, trousseau de   34
Compteur   121
Compteur de la colonie de Québec   121
Condensateur   52
Congélateur   23
Corrosif   3
Couteau à canne de verre   54
Couvercle   57
Couverture anti-feu   1
Crayon à cire   38
Crayon-feûtre   38
Creuset 75, 105
Cuillère   26
Cuve de comptage   117
Cylindre gradué   63

- D -

Danger 2, 3
Débimètre   92
Degré moyen d'humidite   100

Déminéralisateur   109
Densitomètre d'analyse   94
Dessicateur   99
Dialyse   94
Diapositive 116
Disjoncteur   15
Dissection, trousse de   114
Diviseur   98
Douche   1

  - E -

Eau chaude   12
Eau distillée   12
Eau froide   12
Echantilloneur automatique   95
Echelle   68
Egoutteur 125
Electrode   87
Electrophorèse à colonne de gel   94
En panne   3
Entonnoir 104
Entonnoir filtre de Buchner   105
Eprouvette-test   52
Epurateur 106
Escabeau   9
Etiquettes   37
Etiquettes et affiches   2
Etuve 73, 99
Etuve à vide 100
Evaporateur instantané   109
Eviter le contact avec la peau   3
Explosif   2
Extincteur   1
Extracteur de Goldfisch   97

  - F -

Filtre de séparation   95
Fiole   57
Flacon conique d'Erlenmeyer   56
Flacon de compte-gouttes   29
Flacon de distillation   108
Flacon de Florence   57
Flacon laveur   28
Flacon ou plat de coloration   118
Flacon volumétrique   66
Fluoromètre   92
Four à moufle   75

- G -

Gants à jeter   28
Gants de caoutchouc   28
Gants en amiante   28
Gaz   13
Glace sèche, boîte   24
Glace sèche, cuve de stockage de   23
Graisse de bouchon d'arrêt   65

- H -

Haute pression, sous   3
Homogénéiseur   84
Homogénéiseur à emulsion   84
Horloge murale   30
Hotte de ventilation   6
Hotte, utiliser sous   3
Hydromètre   70
Hydroxyde d'ammonium   42
Hydroxyde de calcium   42
Hydroxyde de potasse   42
Hydroxyde de sodium   42

- I -

Incubateur   73
Indicateur   102
Inflammable   3
Installations   11
Interrupteur automatique   30

- K -

Kjeldahl, flacon de   101
Kjeldahl, unité de digestion de   101
Kjeldahl, unité de distillation de   101

- L -

Laine de verre   104
Lampe à alcool   78
Lave-oeil   2
Laveur de pipettes   124
Laveur de verres   124
Limeuse en métal   35

Loupe   116
Lunettes de sécurité   4
Lunettes, mettez vos   3

- N -

Nitrogène (N₂)   13
Nettoyer, produit à   29

- O -

Oxygène (O₂)   13

- P -

Papier filtre   104
Papier indicateur   88
Papier-mouchoir   29
Perceuse   35
pH, appareil de mesure du   87
pH, solution tampon   87
Pilon et mortier   81
Pince   47
Pince avec support   48
Pince de burette   47
Pince de Mohr   51
Pince de tuyau   60
Pinces   26
Pipette au zéro automatique   65
Pipette en seringue   62
Pipette graduée   62
Pipette volumétrique   62
Placard de rangement   6
Plaque chauffante   73
Plat de Petri   120
Poids   69
Poignée de sécurité   53
Poison   3
Polarimètre   71
Pompe   66
Pompe à main   21
Pompe à main, modèle au sol   21
Pompe à vide   20
Pompe à vide et à pression   20
Pompe à vide, mouvement de   20
Presse hydraulique   85
Prise électrique   15

- R -

Raccord et douille 51
Radiations, danger de 3
Réactif 42
Réactif, désignation de 43
Rechargeur de piles 16
Récipient gradué 64
Réfractomètre 71
Réfractomètre à main 71
Réfrigérateur-congélateur 23
Réfrigérateur, garder au 3
Règle 38
Régulateur de pression 18
Risque biologique 2
Robinet 11
Robinet d'arrêt en verre 52
Robinet-mélangeur d'eau 11
Rodage normalisé 51

- S -

Sac en papier 27
Sciseaux 33
Sciseaux pour aluminium 33
Seau 56
Séchage par évaporation de glace 100
Secours, trousse de premier 1
Securité, materiel de 1
Siphon 13
Socle 6
Soudeuse 35
Soupape en aiguille 13
Spatule 26
Spectrophomètre 91
Spectrophomètre, cellule de 91
Stérile 3
Stérilisateur 77
Stérilisateur d'instruments 77
Style à bille 37
Support à tubes d'essais 53
Support d'anneaux 46
Support de bouteille à gaz 18
Support de burette 46
Support de creuset 106
Support de pinces 47
Support d'éprouvette test 53
Support ou cadre 46
Supports 46

- T -

Table   7
Tableau ou carte périodique 40
Table d'opération pour animaux 114
Tablier 27
Tabouret de laboratoire   9
Tamis 95
Téléphone   2
Tenailles 33
Tenailles aux pinces en aiguilles 33
Tenailles coupe-fer 33
Thermique, cordon 74
Thermique, manchon 74
Thermomètre 76
Thermomètre à cadran 76
Thermomètre électronique 76
Thermos, bouteille 24
Tire-bouchon 58
Tissu de lentille 117
Tourne-vis 34
Transformateur variable 15
Trépied 47
Tube à point de fusion 79
Tube capillaire 50
Tube d'alimentation en eau 113
Tube de culture 122
Tuyauterie 60
Tuyauterie en caoutchouc 60
Tuyauterie en plastique 60

- U -

Unité de traitement des tissus 118

- V -

Vapeur 12
Verre de montre 53
Verres 50
Vide 13
Viscomètre 70
Vis d'établis 35
Voltamètre Ohm 88

- A -

Absaugeflasche 105
Abdampfschale 27
Abspeerhahn 12, 65
Abspeerhahnschmiere 65
Abtastdenitometer 94
Abzug 6
Aktenschrank 7
Alkohol 42
Ammoniumhydroxid 42
Ampulle 57
Analytische Waage 68
Anschluß 15
Anschlußkasten 15
Asbesthandschuhe 28
Aspirator (Wasserpumpe) 21
Atemgerät 4
Aulklebezettel 37
Aufspannblock 35
Augenspülwasser 2
Automatische Nullpipette 65
Automatische Vollpipette 63
Automatische Waage 69
Automatischer Probennehmer 95

- B -

Babcock-Flasche 98
Babcock-Zentrifuge 98
Barometer 88
Batterie; Aufladenbar und Hichtaufladenbar 16
Becher 56, 64
Behälter aus Aluminiumfolie 99
Berührung der Haut vermeiden 3
Beschriftungsgerät 37
Bewegung des Pumpenrotors 20
Bohrenspitze 35
Bombenkalorimeter 89
Brenner 77
Brutapparat 73
Büchner-Filtertrichter 105
Bürettengestell 46
Bürettenklammer 47
Bürste 29

- C -

Calciumhydroxid 42
Chromatographiekammer 96

Gasflasche oder Gasbehälter   18
Gefahr   3
Gefahr der biologischen Verseuchung   2
Gefahr der radioaktiven Verstrahlung   3
Gefriertrocknung   100
Gel-Säuleelektrophorese   94
Gewbwschleifmaschine   83
Gewebebearbeitungseinheit   118
Gewichtssatz   69
Gift   3
Glasshahn   52
Glassrohr   50
Glassrohrfräser   54
Glasstab   50
Glasstöpsel   57
Glasswolle   104
Glühtiegel   105
Goldfisch-Extraktor   97
Grosse Schüttelvorrichtung   82
Gummibirne   29
Gummihanschuhe   28
Gummischlauch   60

- H -

Haarrörchen   50
Hahn   11
Hammer   34
Handpumpe   21
Handrefrektometer   71
Handschleifmaschine   83
Handschuhe zum einmaligen Gebrauch   28
Handschutz   53
Hebelwaage   68
Heiß   3
Heißes Wasser   12
Heizband   74
Heizpilz   74
Heizplatte   73
Homogenisierapparat   84
Hydraulische Presse   85
Hydrometer   70

- I -

Im Kühlschrank aufbewahren   3
Impfungsnadel oder Schlinge   122
Intervalstoppuhr   30

Messzylinder 63
Metalfeile 35
Metalschneidende Schere 33
Mikrometer 117
Mikroskop 116
Mikrotom 118
Mixer 82
Molekulmodelle 45
Mörser und Mörserkeule 81
Muffelofen 75
Mühle 83

- N -

Nadelförmige Zange 33
Nadelventil 13
Natriumhydroxid 42
Notizheft 26

- O -

Ofen 73
Operationstisch für Tiere 114
O-Ringgelenk 51

- P -

Papiertücher 29
Periodensystem oder Tabelle 40
Petrischale 120
pH-Meter 87
pH-Pufferlösung 87
Pipettenbirne 63
Plastikschlauch 60
Polarimeter 71
Präpariersatz 114
Pumpe 66
Putzmittel 29
Pyknometer 70

- Q -

Quebec-Kolonienzähler 121
Quetschhahn 51

- R -

Reagenspapier 43
Reagenzglass 52
Reagenzglassgestell 53
Reagenzglasshalter 53
Rechner 38
Refraktometer 71
Reinigungsflasche 28
Registrierapparat 102
Respiratometer 111
Retorte 108
Retortengestell 46
Rührer 26, 81

- S -

Salpetersäure 42
Salzsäure 42
Sauerstoff (O_2) 13
Scheidetrichter 95
Schere 33
Schieber 116
Schlauchhahn 60
Schmelzpunktbestimmungsapparat 79
Schmelzpunktrörchen 79
Schmelztiegel 75
Schrank 7
Schraubenschlüssel (einstellbar) 34
Schraubenzieher 34
Schaufel 26
Schreibtisch 7
Schürze 27
Schüttelvorrichtung 82
Schutzbrille 4
Schwefelsäure 42
Setz die Schutzbrille auf 3
Sieb 95
Siphon 13
Soxhlet-Extraktor 97
Spachtel 26
Spektralphotometer 91
Spektralzelle 91
Spirituslampe 78
Sprengstoff 2
Spritzpipette 62
Spühlbecken 11
Standard Verjüngtes Glasschliff 51
Stativ 46
Staubprobansammler 106
Stechzirkel 98
Steril 3
Sterilisationskasten für Petrischale 121

Sterilisator   77
Sterilisator der Instrumente   77
Stickstoff (N₂)   13
Stoppuhr   31
Stöpsel   57
Stoßverdampfer 109
Streichhölzer   78
Stützbare Handpumpe   21

- T -

Tauchsieder   74
Telefon   2
Tiegalhalter 106
Tisch   7
Thermometer   76
Thermometer mit Zifferblatt   76
Träger   90
Tränkrohr 113
Transformator ("Variac")   15
Trennsäule   96
Trichter 104
Trockeneisbehälter (Kohlendioxidsehneebehälter) .   23
Trockeneiskasten (zur Lagerung und Transport) . .   24
Trockner   99
Trockner für Glass 125
Tropfflasche   29
Tüten   27

- U -

Uhrglass   53
Unter hohem Druck   3
Unter niedrigem Druck   3

- V -

Vakuum   13
Vakuumofen 100
Vakuumpumpe (Drehkolbenölpumpe)   20
Vakuum- und Druckpumpe   20
Verdünnungsflasche 120
Viscozimeter   70
Vollpipette   62
Volt-Ohm-Amperameter   88
Vorsicht   2

- W -

Wanduhr   30
Warm und kalt Wasserhahn   11
Warntafeln, Warnzeichen   2
Waschvorrichtung für Glass   124
Waschvorrichtung für Pipetten   124
Wasserbad   75
Wasserdestillierappart   108
Wichtewaage   69

- Z -

Zählkammer   117
Zähler   121
Zange 26, 33
Zeitschalter   30
Zentrifuge   89
Zentrifugenglass   90
Zentrifugenkopf (Rotor)   90
Zerfressend   3

- A -

alkohol   42
aparat destylacyjny 108
aparat destylacyjny stosowany w metodzie
     Kjeldahla 101
aparat do oznaczania zawartości wody 100
aparat do przygotowywania preparatów
     histologicznych 118
aparat do spalania próbek stosowany w
     metodzie Kjeldahla 101
aparat Goldfischa   97
aparat Soxhleta   97
aparatura pomiarowa   87
aparatura rozdzielcza   94
aparat Warburga 111
areometr   70
autoklaw   77
automat do rotacyjnego przesuwania próbek . . .   95
azot (N_2)   13

- B -

badania mikroskopowe 116
badanie przemian gazowych 111
bagietka szklana   50
barometr   88
bateria   16
bibuła filtracyjna 104
bibułka do czyszczenia 29, 117
biureta wielomiarowa   64
biurko   7
bomba kalorymetryczna   89
bufory wzorcowe   87
butelka   56
butelka do rozcieńczeń 120
butla gazowa   18

- C -

chłodnica   52
chromatografia cienkowarstwowa   96
cieplarka   73
ciężar i skład próbki   68
cylinder miarowy   63
cyrkiel pomiarowy   98
czasomierz, minutnik   30

- D -

demineralizator 109
desintometr skaningowy 94
destylacja 108
destylarka 108
dializator 94
długopis 37
dozownik 63
drabinka 9

- E -

Eksykator 99
Elektrody 87
elektroforeza kolumnowa na żelu 94
etykiety na odczynnikach 43
eza 122

- F -

fartuch ochronny 27
fiolka 57
filtracja 104
fleometr 92
fluorymetr 92
folia 27

- G -

gaśnica 1
gaz 13
gaz w butlach 18
gilza ekstrakcyjna 97
gilzy do probówek wirówkowych 90
głowica wirówki 90
gniazdko elektryczne 15
gruszka gumowa 29
grzejnik zanurzeniowy 74

- H -

homogenizator 84
homogenizator emulsyjny 84
homogenizator szklany do miękkich tkanek 83

- I -

imadło   35

- K -

kalkulator   38
karmidełko   113
klamra zaciskowa   51
klatka dla zwierząt   113
klucz francuski   34
klucz płaski do nakrętek   34
koc przeciwpożarowy   1
kolba destylacyjna   108
kolba Kjeldahla   101
kolba miarowa   66
kolba płaskodenna   57
kolba ssawkowa   105
kolba stożkowa   56
kolumna chromatograficzna   96
komora chromatograficzna   96
komora do liczenia krwinek i drobnoustrojów . .   117
konieczna odpowiednia wentylacja   3
korek   57
korek szklany   57
korek z kory drzewa   58
korek z otworem do wprowadzania lejka   106
korkobór   58
kran wodociągowy   11
kran wodociągowy z zimną i ciepłą wodą   11
kroplomierz   29
krzesło laboratoryjne   9
kurek 12, 65
kurek szklany   52
kuweta   91
kwas azotowy   42
kwas octowy   42
kwas siarkowy   42
kwas solny   42

- L -

lampka spirytusowa   78
lejek   104
lejek Büchnera   105
lejek rozdzielczy   95
licznik   121
licznik kolonii   121
linijka   38
lodówka z zamrażalnikiem   23
lupa   116
lutownica   35

- Ł -

łapa statywu   47
łapa statywu z łącznikiem   48
łaźnia wodna   75
łącznik   47
łącznik pierścieniowy   51

- M -

materiał wybuchowy   2
meble laboratoryjne   6
menzurka 64
miernik uniwersalny 88
mieszadło 26, 81
mieszadło magnetyczne 81
mikrobiologia 120
mikrometr 117
mikroskop 116
mikrotom 118
mikser 82
młotek 34
młynek 83
modele molekularne 45
moździerz i pistol 81

- N -

naczynia i ich zamknięcia 56
naczynie do batwienia 118
naczynko z folii aluminiowej 99
nalepka 37
napisy i znaki ostrzegaiące   2
niebezpieczeństwo   3
niebezpieczeństwo skażenia biologicznego 2
notatnik 26
nożyce do cięcia blachy 33
nożyczki 33

- O -

ochraniacz dłoni 53
odczynniki chemiczne 42
ogrzewanie 73
okulary ochronne   4
ołówek woskowy, dermatograf 38
ostrożnie   2

- P -

palnik   77
papierki wskaźnikowe   88
para wodna   12
parownica   27
pH-metr   87
piec muflowy   75
piknometr   70
pilnik   35
pipeta automatyczna jednomiarowa   65
pipeta miarowa   62
pipeta napełniana gruszką gumową   63
pipeta strzykawkowa   62
pipeta wielomiarowa   62
pisak, flamaster   38
płaszcz grzejny   74
płuczka   106
płuczka do pipet   124
płytka Petriego   120
podgrzewacz   73
podłoża   120
pod niskim ciśnieniem   3
podręczne narzędzia laboratoryjne   33
podręczny sprzęt laboratoryjny   26
pod wysokim ciśnieniem   3
poidełko   113
pojemnik do przechowywania suchego lodu   23
pojemnik do sterylizacji płytek Petriego   121
polarymetr   71
połączenie na standardowy szlif stożkowy . . .   51
połączenie na szlif kulisty   51
pompa próżniowa   20
pompa próżniowo-ciśnieniowa   20
pompka   66
pompka ręczna   21
pompka ręczna (opierana o podłogę)   21
pompka wodna   21
pompy próżniowe i powietrzne   20
powietrze   12
prasa hydrauliczna   85
probówka   52
probówka bakteriologiczna   122
probówka do oznaczania temperatury topnienia   79
probówka wirówkowa   90
próżnia   13
prysznic   1
przechowywać w lodówce   3
przechowywanie w niskich temperaturach   23
przenośne meble laboratoryjne   9
przybory biurowe   37
przyrząd do oznaczania temperatury topnienia   79
przyrząd do rozwijania i cięcia taśmy klejącej   37
przyrząd do tłoczenia napisów na plastikowych
     nalepkach   37
pudło do transportu i przechowywania suchego
     lodu   24

- Ś -

środek zmywający   29
śróbokręt   34

- T -

tabela układu okresowego pierwiastków   40
taśma grzejna   74
telefon   2
termometr   76
termometr elektroniczny   76
termometr tarczowy   76
termos   24
tlen (O_2)   13
tłuszczomierz Babcocka   98
torebka   27
transformator   15
trójnóg   47
trucizna   3
tryskawka   28
tygiel 75, 105

- U -

uchwyt do biuret   47
uchwyt do mocowania butli gazowej   18
uchwyt do probówek   53
unikać kontaktu ze skórą   3
urządzenia elektryczne   15
urządzenie do cięcia szklanych rurek   54
urządzenie do ładowania baterii   16
uszkodzone   3

- W -

waga analityczna   68
waga automatyczna jednoszalkowa   69
waga do określania ciężaru właściwego   69
waga sprężynowa   68
waga trójramienna z przesuwnymi odważnikami .   68
wata szklana 104
wiadro   56
wieczko, nakrętka   57
wiertło   35
wieszak do suszenia szkła laboratoryjnego . . . 125
wirówka   89
wirówka Babcocka   98
wiskozymetr   70
woda destylowana   12
woda do płukania oczu   2

woda gorąca   12
woda zimna   12
wodorotlenek amonowy   42
wodorotlenek potasowy   42
wodorotlenek sodowy   42
wodorotlenek wapniowy   42
wstrząsarka   82
wyciąg   6
wyłącznik zegarowy   30
wyparka 109
wyposażenie   11
wyposażenie związane z bezpieczeństwem pracy .   1
wysoka szafka laboratoryjna   6
wysoka szafka laboratorujna z przesuwanymi
    drzwiami   7
wysoka temperatura   3
wziewnik   4

- Z -

zaciskacz do rurek   60
zagrożenie pożarowe   3
zagrożenie promieniowaniem   3
załóż okulary ochronne   3
zapalacz do gazu   78
zapałki   78
zasada żrąca   3
zawór iglicowy   13
zegar ścienny   30
zestaw kluczy wpustowych sześciokątnych   34
zestaw odważników   69
zestaw pierwszej pomocy medycznej   1
zestaw sekcyjny 114
zlew   11
zlewka 56, 64
zmywanie 124
zmywarka do szkła laboratoryjnego 124

- A -

Acido Acético 42
Acido Clorhídrico 42
Acido Nítrico 42
Acido Sulfúrico 42
Agitador 81
Agitador magnético 81
Agitador-Vibrador 82
Agua caliente 12
Agua destilada 12
Agua fría 12
Aire 12
Alcohol 42
Alicates 33
Alicates de punta fina 33
Archivador 7
Arcón para hielo seco 23
Areómetro 70
Armario 6
Asa de platino 122
Autoclave 77

- B -

Balanza analítica 68
Balanza digital 69
Balanza para densidades 69
Balanza para humedad 100
Baño maría 75
Baño para punto de fusión 79
Barómetro 88
Báscula 68
Batería 16
Batidora 82
Bebedero 113
Bolígrafo 37
Bolsa 27
Bomba 66
Bomba calorimétrica 88
Bomba de vacío 20
Bomba de vacío de Agua 21
Bomba de vacío y compresor 20
Bomba manual 21
Bomba Peristáltica 20
Botella 56
Botiquín 1
Broca 35
Bureta 64

- C -

Cabezal de centrífuga 90
Cabezal múltiple de centrífuga 90
Caja de conexiones electricas 15
Calculadora 38
Caliente 3
Cámara cromatográfica 96
Cámara cuentaglóbulos 117
Cápsula 27
Cargador de baterías 16
Cartucho de extracción 97
Cáustico 3
Centrífuga 88
Centrífuga de Babcock 98
Cilindro de gas 18
Cilindros para esterilizar placas 121
Cinta Adhesiva 37
Cinta calefactora 74
Columna cromatográfica 96
Comedero 113
Compás de puntas 98
Condensador 52
Contador 121
Copa graduada 64
Corcho 58
Corrosivo 3
Cortadora de tubos de vidrio 54
Criba 95
Crisol 75
Crisol de Gooch 105
Cromatografía en capa fina 96
Cronómetro 31
Cubeta de tinción 118
Cubeta para espectrofotometría 91
Cubo con asa 56
Cuchara de agitación 26
Cucharón 26
Cuenta colonias 121

- D -

Delantal 27
Densitómetro 94
Desecador 99
Desmineralizador 109
Destilación, aparato para 108
Destilación, matraz de 108
Destilador de agua 108
Destornillador 34
Diálisis, equipo de 94
Disección, equipo de 114
Ducha 1

- E -

Electrodo 87
Electoforesis en gel, equipo para 94
Embudo 104
Embudo de Buchner 105
Embudos Buchner, conos de goma para 106
Embudo de separación 95
Emulsificador 84
Encendedor 78
Enchufe 15
Equipo de inclusión 118
Equipo de muestreo automático 95
Equipo de seguridad 1
Escalera 9
Escobillón 29
Escurridor 125
Espátula 26
Espectrofluorímetro 92
Espectrofotómetro 91
Estantería 7
Estéril 3
Esterilizador de instrumentos 77
Estufa 73
Estufa de desecación 99
Estufa de vacío 100
Etiqueta 37
Etiquetado de Reactivos 43
Etiquetador 37
Etiquetas y señales 2
Evaporador 109
Explosivo 2
Extintor de incendios 1
Extractor de Goldfisch 97
Extractor de Soxhlet 97

- F -

Flotámetro 92
Fósforos de seguridad 78
Frasco cuentagotas 29
Frasco de dilución 120
Frasco lavador 28
Frasco lavador de gases 106
Fregadero 11
Frigorífico-Congelador 23
Funciona, no 3

- G -

Gafas protectoras   4
Gafas, use   3
Gas 13
Goma, tubo de 60
Gradilla 53
Granatario 68
Grasa para buretas 65
Grifo 11, 12
Grifo combinado 11
Guantes de amianto 28
Guantes de goma 28
Guantes desechables 28

- H -

Hidróxido Amónico 42
Hidróxido Cálcico 42
Hidróxido Potásico 42
Hidróxido Sódico 42
Hoja de aluminio 27
Homogeneizador 84
Homogeneizador de vidrio 83
Horno 73
Horno de mufla 75

- I -

Inflamable   3

- J -

Jaula 113
Jeringuilla micrométrica 62
Juego de pesas 69

- K -

Kjeldahl, equipo de destilación 101
Kjeldahl, equipo de digestión 101
Kjeldahl, matraz de 101

- L -

Laboratorio, carro de   9
Lámpara de alcohol   78
Lana de vidrio   104
Lapiz de cera   38
Lavadora   124
Lavador de pipetas   124
Lavaojos   2
Libro de notas   26
Lima   35
Limpiador   29
Liofilizador   100
Lupa   116

- LL -

Llave Ajustable   34
Llave de dos bocas   34
Llave de paso de vidrio   52
Llave de vidrio   65
Llaves para tuercas hexagonales   34

- M -

Mango de seguridad   53
Manta calefactora   74
Manta contra incendios   1
Mantener en refrigeración   3
Martillo   34
Mascarilla   4
Matraz de Babcock   98
Matraz de fondo plano   57
Matraz erlenmeyer   56
Matraz Kitasato   105
Matraz volumétrico   66
Mechero   77
Medio de cultivo   120
Mesa   7
Mesa de despacho   7
Mesa de operaciones   114
Micrómetro   117
Microscopic   116
Microtomo   118
Modelos moleculares   45
Molino   83
Mortero y mano   81

- N -

Nitrógeno   13
Nuez   47

- O -

Oxígeno   13

- P -

Pañuelo   29
Papel de filtro 104
Papel indicador   88
Papel limpialentes 117
Peligro   3
Pera   63
Pera de goma   29
pH-Metro   87
Picadora manual   83
Picnómetro   70
Piel, evite el contacto con la   3
Pinza   47
Pinza con nuez   48
Pinza de mohr   60
Pinza de sujeción para cilindros de gas   18
Pinza para buretas   47
Pinzas para crisoles   26
Pinza para esmerilados de rótula   51
Pinzas para tubos de ensayo   53
Pipeta aforada   62
Pipeta automática 63, 65
Pipeta graduada   62
Placa calentadora   73
Placa de Petri 120
Platillo de hoja de aluminio   99
Polarímetro   71
Portaobjetos 116
Precaución   2
Prensa hidráulica   85
Presión elevada   3
Presión reducida   3
Presión, regulador de   18
Probeta   63

- R -

Recipiente para hielo seco   24
Refractómetro   71
Refractómetro manual   71
Registrador 102
Regla   38
Reloj avisador   30
Reloj de pared   30
Respirómetro 111
Riesgo Biológico   2
Riesgo Radiactivo   3
Rotulador   38

- S -

Sifón   13
Silla o taburete de laboratorio   9
Soldador   35
Solución tampón   87
Soporte de aro   46
Soporte o Emparrillado   46
Soporte para buretas   46
Sustentacion, módulo de   6

- T -

Taladratapones   58
Tapa   57
Tapón   57
Tapón de vidrio   57
Teléfono   2
Tenazas   33
Termo   24
Termómetro   76
Termómetro electrónico   76
Termómetro industrial   76
Termostato   74
Tijeras   33
Tornillo de banco   35
Transformador   15
Trípode   47
Tubo capilar   50
Tubo de centrífuga   90
Tubo de cultivo 122
Tubo de ensayo   52
Tubo de plástico   60
Tubo de vidrio   50
Tubo para punto de fusión   79

- U -

- V -

HOW DO YOU SAY "HELP" IN ENGLISH ?

- A -

ACS 43
Açma kapama düğmesi 1, 121
Agırlıklar 68
Agırlık kontrol düğmesi 68
Ağırlık seti 69, 72
Ağız 56, 63
Akış kontrolu 13
Akrep ve yelkovan 30, 31
Alçak basınç 3, 5
Alevlenebilir 3, 5
Aliminyum foil 27, 32, 99, 103
Aliminyum foil kapak 27
Aliminyum kuru madde kabı 27, 99, 103
Alkol 42, 44
Alkol veya gazyağı lambası 78, 80
Ameliyat 114, 115
Ameliyat takımı 114, 115
Ameliyat masası 114, 115
Amonyum hidroksit 42, 44
Ampermetre 16, 88, 93
Anahtar 34, 36
Anahtar takımı 34, 36
Araba 9, 10, 18
Arka 9
Artery forceps 114
Aşindirici (corrosive) 3, 5
Asbest eldiven 28, 32
Asetik asit 42, 44
Asit 42, 44
Aspiratör 21, 22, 101
Atom 45
AVO-metre 88, 93
Ayak 7
Ayak pompası 21
Ayar düğmesi 34
Ayarlama çizgisi 66
Ayırma alet ve cihazları 94, 103
Ayırma hunisi 95, 103
Azot (N_2) 13, 14

- B -

Babkok santrifüjü 98, 103
Babkok şişesi 98, 103
Bağ 45
Bağlantı 51, 55
Banyo 75, 78, 80, 118
Barometre 88
Basınç ayarlayıcısı 18, 19
Basınç göstergesi 18, 20, 21
Basınç kontrol vanası 18

Başparmak 28
Battaniye 15
Baz 44
Beherglas 56, 59, 64, 67
Besiyeri 120, 123
Bıçak 118
Bilezikli dayama ayağı 46, 47, 49
Bir çeşit süzgeç 105
Bir çeşit elcek (cam çubuk için) 53, 55
Biyolojik tehlike 2, 5
Blender 82, 86
Bomb 89
Bomb kalorimetresi 89, 93
Boncuk 9
Boru veya hortum 50, 54, 55, 60
                                             61
Boyama malzemesi 118, 119
Bozuk 3, 5
Buffer çözeltisi 87, 93
Buhar 12, 14
Bulaşık eldiveni 28, 32
Bulaşık makinası 124, 126
Bulb 29, 32, 63, 67
Bunzen 77, 80, 108
Bühner süzgeci 105, 107
Büret 64, 67
Büret ayaklığı 46
Büret kıskacı 47, 49
Büro malzemesi 37, 39
Büro tipi dosya dolabı 7, 8
Büyüteç 116, 119, 121
Buzdolabı 3, 5, 23, 25
Buzdolabı ve derin dondurucu 23
Buzdolabında muhafaza et 3, 5

- C -

Cam 50, 55
Cama yazar kalem 38, 39
Cam balon 57, 59
Cam boru 50, 55
Cam boru kesicisi 54, 55
Cam çubuk 50, 55
Cam elektrot 87
Cam musluğu 65, 67
Cam şişe 56
Cam tıpa 57, 59
Cam vana 52, 55
Cam yünü 104, 107
Ca(OH)$_2$ 42
Cerrah mili 114
Cetvel 38, 39, 114

CH₃CH₂OH 42
CH₃COOH 42
Checknut 48
Cımbız 69, 114
Clip 37, 65
Concentric rings 75
Contalı ekleme 51, 55
CW 12

- Ç -

Çakmak 78, 80
Çalışma ünitesi 6, 8, 11
Çalkalayıcı 82, 86
Çeker ocak 3, 5, 6, 8
Çeker ocagı kullan 3, 5
Çekiç 34, 36
Çekmece 6
Çerçeve 46, 49
Çevirme kolu (krank) 89, 118
Çıkış 15, 17, 20
Çıkış borusu 21
Çoklu priz 15, 17
Çubuk 50, 55

- D -

Dalga boyu ayar dügmesi 91
Damlalık 29, 32
Dayama (destek) 46, 49
Dayama çerçevesi 46
Değirmen 83, 86
Delik 57
Deney hayvanları için özel aletler 113
Deney tüpü maşası 53, 55
Deney tüpü . . . . 27, 52, 53, 5!
Densitometre 94, 103
Dereceli beherglas 64
Dereceli pipet 62, 67
Deri temasindan sakın 3, 5
Desikatör 99, 103
Desikatör yagı (gres) 65, 67
Dial (gösterge) 2, 68, 76, 80
Distilasyon 108, 110
Distilasyon balonu 108, 110
Dilüsyon şişesi 120, 123
Dişli musluk 12
Distile su 12, 14
Diyaliz cihazı 94, 103
Doku işleme ünitesi 118, 119
Doku öğütücüsü 83, 86

Dolap 3, 5, 6, 7
                                             8, 96
Dolum çizgisi 120
Dondurarak kurutma 100, 103
Dondurucu 23, 25
Duş için çeşme 1, 5
Duvar saati 30, 32
Düğme 15
DW 12

- E -

Eğe 35, 36
Ekstraktör 97, 103
Ekstraksiyon tüpü 97, 103
Elcek 7, 21, 26, 29
                                             34, 53, 55, 56
                                             83, 85
Eldiven 28, 32
Elek 95, 103
Elektrik 15, 17
Elektrikli havya 35, 36
Elektroforez 94, 103
Elektronik termometre 76, 80
Elektrot 87, 93
El değirmeni 83, 86
El pompası 21, 22
El refraktometresi 71, 72
Emme 63
Emniyet alet ve ekipmanları 1, 5
Emniyet bilgileri 43
Emniyet gözlüğü 3, 4, 5
Emülsiyon homojenizatörü 84, 86
Erime noktası tayini için banyo 79, 80
Erime noktası tayin tüpü 79, 80
Erlenmayer 56, 59
Eter ekstraksiyon cihazı 97, 103
Etiket 2, 5, 37, 39
                                             43, 44
Etiket yazıcısı 37, 39
Etiket yeri 120
Evaporatör 109, 110
Ezme, parçalama ve karıştırma aletleri . . 81

- F -

Fırın 73, 80, 99, 100
                                             103
Fırça 29, 32
Fiş 15, 16, 74
Fitil 78
Flaş evaporatör 109, 110

Florometre 92, 93
Flowmetre 92, 93
Fren 98

- G -

Garantili 43
Gaz 13, 14, 18, 19
                                          78, 80
Gaz ayarı 77
Gaz ölçümü 111, 112
Gaz tüpü 18
Gaz tüpü emniyet kemeri 18, 19
Giriş 20
Giriş borusu 13, 64, 65, 124
Gösterge 18, 20, 21, 85
Göstergeli termometre 76, 80
Göz yıkamak için çeşme 2, 5

- H -

Hacim pipeti 62, 67
Hacimli balon 66, 67
Hassas terazi 68, 72, 100, 103
Hava 12, 14, 20, 22
Hava ayarı 77
Hava filtresi 20
Havagazı 13
Havan ve tokmak 81, 86
Hazne 118
HCl 42
Hesap makinası 38, 39
Hidroklorik asit 42, 44
Hız 81, 89
Hız kontrolu 82, 89
Hız göstergesi 89
Hidrolik pres 85, 86
Hidroksit 42, 44
Hidrometre 70, 72
HNO_3 42
Homojenizatör 84, 86
Hortum bağlantısı 18
Hot plate 73, 80
Hot plate yükseklik ayarı 97
H_2SO_4 42
Huni 95, 103, 104
                                          105, 107
Hücre 91, 93
HW 12

Isı kuşağı 74, 80
Isı mantosu 74, 80
Isıtıcı 101
Isıtma 73, 80
Işık kaynağı 116

- İ -

İçecek tüpü 113, 115
İlk yardim kutusu 1, 5
İmbik (damitıcı) 108, 110
İmpinger 106, 107
İnce ayar 116
İnce tabaka kromotografisi 96, 103
İnkübatör 73, 80
İşaret 2, 5

- J -

Jel elektroforezi 94, 103

- K -

Kaba ayar 116
Kaba dereceli silindir 64, 67
Kafes 113, 115
Kağıt mendil 29, 32
Kalem 37, 38, 39
Kalın uçlu kalem 37, 39
Kalsiyum hidroksit 42, 44
Kanca 114
Kap 27, 32, 99, 103
                                          113, 115, 118, 119
                                          120, 121, 123
Kap tipi malzemeler 56, 59
Kapak 18, 29, 57, 59
                                          78, 122
Kapak ayar kolu 77
Kapaklı deney tüpü 122, 123
Kapı 6, 23
Kapillar cam boru 50, 55
Karbondioksit 23, 24, 25
Karga burnu 33
Kariştırıcı 81, 86
Kasnak 83, 89
Katalog numarası 43
Kavanoz 118, 119
Kayış 4, 18, 83, 89
Kaynatıcı 77, 80
Kazıyıcı (spatula) 26, 32
Kefe 68, 69
Keldal balonu 101, 103

Keldal distilasyon ünitesi 101, 103
Keldal yakma ünitesi 101, 103
Kelebek vida 47, 48
Kepce 26, 32
Kesici 37, 54, 55
Kilit manivelası 7
Kimyasal madde 42, 44
Kimyasal madde etiketi 43, 44
Kiprit 78, 80
Kıskaç 18, 19, 47, 48
                                            49, 51, 55, 60
                                            61, 83
Kıskaç tutucu 47, 49
Kıskaçlı büret tutucu 48, 49
KOH   42
Kolon 96, 103
Koloni sayıcısı 121, 123
Konik cam kap 56, 59
Kontrol lambası 15
Koruyucu gözlüğünü tak 3, 5
Kostik 3, 5
Kova 56, 59
Kromotografi 96, 103
Kromotografi kabineti 96
Kromotografi kolonu 96, 103
Kronometre 31, 32
Kulp   56
Kül kabı (crucible) 75, 80, 105, 107
Kullanılıp atılabilen eldiven 28, 32
Kurbağacık 34, 36
Kuru buz banyosu 100
Kuru buz (CO$_2$) dolabı 23, 25
Kuru buz kutusu 24, 25
Kurutma dolabı 99, 103
Kurutma fırını 73
Kurutma kabı 27, 32
Kurutma rafı 125, 126
Kutu 24, 25, 78, 121
                                            123
Kutup 16

- L -

Laboratuvar donanımı 6, 9
Laboratuvar el aletleri 26, 32, 33
Laboratuvar tipi el arabası 9, 10, 18, 19
Laboratuvar tipi sandalye 9, 10
Laboratuvar tipi saat 30, 32
Lam 116, 119
Lamba 78, 80
Lastik boru kıskacı 60, 61
Lastik bulb 29, 32

Lastik hortum 60, 61
Lastik tıpa 57
Lavoba 11, 14
Lens mendili 117, 119

- M -

Magnetik karıştırıcı 81, 86
Makas 33, 36, 114
Malzeme dolabı 6, 7, 8
Manivela kolu 2, 30
Mantar tıpa 58, 59
Mantar delici 58, 59
Masa 7, 8, 114, 115
Maşa 26, 32
Maşalı kıskaç 51, 55
Matkap (kalem) seti 35, 36
McLead vakum ölçer 100
Mendil 29, 32, 117, 119
Mengene 35, 36
Merdiven 9, 10
Metrik 43
Mikrobiyoloji 120, 123
Mikrometre 117, 119
Mikroskop 116, 119
Mikrotom (kesit alma cihazı) 118, 119
Miktar 43
Mineral uzaklaştırıcı 109, 110
Model 45
Molekül modelleri 45
Motor 20, 83, 89
Musluk 11, 12, 14

- N -

NaOH 42
Naylon torba 27, 32
Nem ölçme terazisi 100, 103
Neşter 114
NH₄OH 42
NIT 13
Nitrik asit 42, 44
Not defteri 26, 32

- O -

Objektif 116
Ofis tipi çalışma masası 7
Ohm 88, 93
Oksijen (O₂) 13, 14
Oküler 116
OSHA 43

Otoklav 77, 80, 121, 123
Otomatik büret 65, 67
Otomatik örnek alıcı 95, 103
Otomatik pipet 63, 67
Oksi 13

- ö -

Ölçü alet ve cihazları 87, 93
Ölçü silindiri 63, 67
Önlük 27, 32
Örnek alıcı 95, 103
Örnek koyma yeri 98
Öze ve iğne 122, 123
Özgül ağırlık terazisi 69, 72

- P -

Parmak 28
Patlayıcı (explosive) 2, 5
Pense 33, 36
Pergel 98, 103
Periyodik cetvel 40, 41
Petri kutusu 120, 123
Petri kutusu sterilizasyon kabı 121, 123
PH 87, 88, 93
pH metre 87, 93
Piknometre 70, 72
Pil (şarz olabilir veya olamaz tip) 16, 17
Pil şarjer (şarz aleti) 16, 17
Pipet 29, 32, 62, 63
                                              65, 67, 126
Pipet emicisi (zararlı sıvılar için) 63, 67
Pipetlik (yıkama için) 124, 126
Plastik hortum 60, 61
Plastik su şişesi 28, 32
Polarimetre 71, 72
Pompa 20, 22, 66, 67
Pompa hareket vanası 20
Porselen kap 27
Potasyum hidroksit 42, 44
Priz 15, 17
Punch 58

- R -

Rack 124, 125, 126
Radyasyon tehlikesi 3, 5
Raf 6
Referans elektrot 87
Refraktometre 71, 72

Regülatör 18, 19
Respirometre 111, 112

- S -

S (es) boru 11
Saat 30, 32
Saat camı 53, 55
Saat kolu 31
Sabit donanım 11
Safiyet tablosu 43
Saf su cihazı 108
Sandalye 9
Santrifüj 89, 93, 98, 103
Santrifüj başı (rotoru) 90, 93
Santrifüj tüpü 90, 93
Sayaç 121, 123
Sayım haznesi. 117
Scanning densitometre 94, 103
Seloteyp (bant) 37, 39
Seri numarası 43
Sıfırlama düğmesi 121
Sivı hacim ölçme malzemeleri 62
Sıcak 3, 5
Sıcaklık kontrol düğmesi 73, 97
Sıcaklık kontrol ünitesi 75
Sıcak soğuk karıştırma musluğu 11, 14
Sıcak su 12, 14
Sifon 13, 14
Silindir 18, 19, 63, 67
Silindir coupling 18
Sodyum hidroksit 42, 44
Soğuk depolama 23, 25
Soğuk su 12, 14, 52, 108
Soğutucu 52
Soksilet cihazı 97, 103
Spatula 26, 32
Spektrofotometre 91, 93
Spektrofotometre tüpü 91, 93
Standart traşlı cam 51
Standart traşlı cam ekleme 51, 55
Steril 3, 5
STM 12
Su 1, 12, 14, 52
                                          108
Su banyosu 75, 80
Su basıncı 21
Su çıkışı 21
Su ısıtıcısı 74, 80
Su trompu 21, 22
Sülfürik asit 42, 44
Süzgeç desteği 106, 107

Süzgeç kağıdı 104, 107
Süzme 104, 105, 107
Süzme erleni 105, 107

- Ş -

Şarz aleti 16, 17
Şirınga tipi pipet 62, 67
Şişe 28, 32, 56, 59
                                         70, 72, 98, 103
                                         120, 123

- T -

Tahliye borusu 124
Tapa 52
Tappet (tripper) 30
Tartım ve örnek kompozisyonu 68
Taşıma 11, 13
Taşıma kabı 64, 65
Taşınabilir laboratuvar donanımı 9
Taşıyıcı 90, 93
Tedbirli ol 2, 5
Tehlike etiket ve işaretleri 2, 5
Tehlike 3, 5
Tekerlek 9
Tel keser (pense) 33, 36
Telefone 2, 5
Temizlik malzemesi (sabun vb) 29, 32
Teneffüs cihazı 4, 5
Teneke makası 33, 36
Terazi 68, 72
Termometre 76, 80, 89, 99
Termostat 74
Termus 24, 25
Tetik veya zemberek 1, 35
Tıpa 57, 59
TLC 96, 103
Top ve yuva şeklinde ek (cam malzelerde) . . 51, 55
Toplama kabı 108
Tornavida 34, 36
Tornavida tipi anahtar takımı 34, 36
Transformotor 15, 17
Tulumba 62
Tüp 79, 80, 90, 93
                                         113, 115, 122, 123
Tüp (gaz) 18, 19
Tüp sehpası 53, 55, 122
Turnusol kağıdı 88, 93

- U -

Uç 35, 37

Ufak şişe 57, 59

- Ü -

Üç ayak (lı ızgara) 47, 49
Üç kollu terazi 68, 72
Üst 23, 24
Üstten kefeli hassas terazi 69, 72

- V -

VAC 13
Vakum 13, 14, 20, 21
                                            22, 106
Vakum girişi 20
Vakum göstergesi 20, 73, 100
Vakum kurutma dolabı 100, 103
Vakum pompası 20, 22, 100
Vakum ve basınç pompası 20, 22
Vana 13, 14, 18, 20
                                            52, 55, 65, 66
                                            67
Viskozimetre 70, 72
Voltaj ayar düğmesi 15
Volt metre 88, 93

- Y -

Yağ haznesi 20
Yağ seviyesi 20
Yakma fırını 75, 80
Yangın örtüsü (battaniye) 1, 5
Yangın söndürücü 1, 5
Yan kol 79, 105, 108
Yazıcı 102, 103
Yelkovan 31
Yeri (kaynağı) 43
Yıkama 124, 126
Yiyecek kabı 113, 115
Yoğunlaştırıcı 52, 55, 97, 100
                                            101, 108
Yüksek basınç 3, 5

- Z -

Zaman 30
Zamanlayıcı 30, 32
Zehir 3, 5
Zincir 114